Annual Review of Broadband Communications

IEC
Chicago, Illinois

About the International Engineering Consortium

The International Engineering Consortium (IEC) is a non-profit organization dedicated to catalyzing technology and business progress worldwide in a range of high technology industries and their university communities. Since 1944, the IEC has provided high-quality educational opportunities for industry professionals, academics, and students. In conjunction with industry-leading companies, the IEC has developed an extensive, free on-line educational program. The IEC conducts industry-university programs that have substantial impact on curricula. It also conducts research and develops publications, conferences, and technological exhibits that address major opportunities and challenges of the information age. More than 70 leading high-technology universities are IEC affiliates, and the IEC handles the affairs of the Electrical and Computer Engineering Department Heads Association and Eta Kappa Nu, the honor society for electrical and computer engineers. The IEC also manages the activities of the Enterprise Communications Consortium.

Other Quality Publications from the International Engineering Consortium

- *Achieving the Triple Play: Technologies and Business Models for Success*
- *Business Models and Drivers for Next-Generation IMS Services*
- *Delivering the Promise of IPTV*
- *Evolving the Access Network*
- *The Basics of IPTV*
- *The Basics of Satellite Communications, Second Edition*
- *The Basics of Telecommunications, Fifth Edition*

For more information on any of these titles, please contact the IEC publications department at +1-312-559-3730 (phone), +1-312-559-4111 (fax), *publications@iec.org*, or via our Web site (http://www.iec.org).

ISBN: 978-1-931695-71-8

International Engineering Consortium
300 West Adams Street, Suite 1210
Chicago, Illinois 60606-5114 USA
+1-312-559-3730 phone
+1-312-559-4111 fax

Contents

Academic Perspectives

Executive Perspectives

Ethernet

Triple Play

Contents by Author

University Program Sponsors

The IEC's University Program, which provides grants for full-time faculty members and their students to attend IEC Forums, is made possible through the generous contributions of its Corporate Members. For more information on Corporate Membership or the University Program, please call +1-312-559-4625 or send an e-mail to *cmp@iec.org*.

Based on knowledge gained at IEC Forums, professors create and update university courses and improve laboratories. Students directly benefit from these advances in university curricula. Since its inception in 1984, the University Program has enhanced the education of more than 500,000 students worldwide.

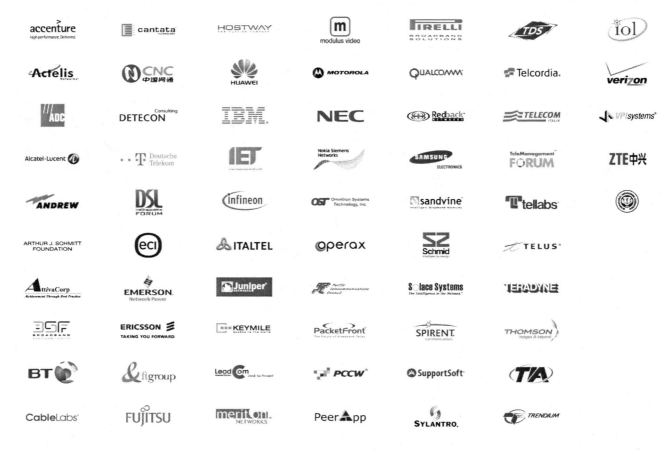

IEC–Affiliated Universities

The University of Arizona
Arizona State University
Auburn University
University of California at Berkeley
University of California, Davis
University of California, Santa Barbara
Carnegie Mellon University
Case Western Reserve University
Clemson University
University of Colorado at Boulder
Columbia University
Cornell University
Drexel University
École Nationale Supérieure des Télécommunications de Bretagne
École Nationale Supérieure des Télécommunications de Paris
École Supérieure d'Électricité
University of Edinburgh
University of Florida
Georgia Institute of Technology

University of Glasgow
Howard University
Illinois Institute of Technology
University of Illinois at Chicago
University of Illinois at Urbana-Champaign
Imperial College of Science, Technology and Medicine
Institut National Polytechnique de Grenoble
Instituto Tecnológico y de Estudios Superiores de Monterrey
Iowa State University
KAIST
The University of Kansas
University of Kentucky
Lehigh University
University College London
Marquette University
University of Maryland at College Park
Massachusetts Institute of Technology
University of Massachusetts

McGill University
Michigan State University
The University of Michigan
University of Minnesota
Mississippi State University
The University of Mississippi
University of Missouri-Columbia
University of Missouri-Rolla
Technische Universität München
Universidad Nacional Autónoma de México
North Carolina State University at Raleigh
Northwestern University
University of Notre Dame
The Ohio State University
Oklahoma State University
The University of Oklahoma
Oregon State University
Université d'Ottawa
The Pennsylvania State University

University of Pennsylvania
University of Pittsburgh
Polytechnic University
Purdue University
The Queen's University of Belfast
Rensselaer Polytechnic Institute
University of Southampton
University of Southern California
Stanford University
Syracuse University
University of Tennessee, Knoxville
Texas A&M University
The University of Texas at Austin
University of Toronto
VA Polytechnic Institute and State University
University of Virginia
University of Washington
University of Wisconsin-Madison
Worcester Polytechnic Institute

Academic Perspectives

An Aspect-Oriented Approach for Dynamic Monitoring of a Service Logic Execution Environment

Paolo Falcarin

Post-Doctoral Fellow, Department of Control and Computer Engineering
Politecnico di Torino, Italy

Laurent Walter Goix

System Engineer
Telecom Italia

Abstract

Service creation environments play a relevant role in new telecom applications because they enable openness and programmability by offering frameworks for the development of value-added services. The Java application programming interface for integrated networks service logic execution environment (JSLEE) specification defines a Java framework for executing event-based distributed services made up of components called service building blocks.

In such a complex architecture, monitoring is an indispensable technique to test the dynamic behavior of a system, debug the code, gather usage statistics, or measure the quality of service. Program instrumentation is needed to insert monitoring codes into the system to be monitored, which is typically a manual and time-consuming task.

This paper describes a language-based approach to automate program instrumentation and monitoring management using a dynamic aspect-oriented programming (AOP) framework.

The basic notions of AOP and the use of the JBoss AOP framework features are described, in order to allow a highly modular and easily configurable implementation of reusable monitoring code. Using an Eclipse-based system administration console, it is possible to manage remotely the dynamic deployment and update of monitoring code in a service deployed on a JSLEE container.

Introduction

A service creation environment addresses the main feature of service programmability—the ability to implement new services faster, with higher software reuse and rapid configuration. Another important issue is the capability to offer to users the same service everywhere, providing a seamless access from different terminals (i.e., mobile phones, SIP-phones [6], and UMTS phones). Among different service creation technologies [1, 2], the Java application programming interface for integrated networks (JAIN) brings service portability, convergence, and secure network access to telephony and data networks.

By providing a new level of abstraction and associated Java interfaces for service creation across public switched telephone networks (PSTN), packet or wireless networks, JAIN technology enables the integration of Internet and telecommunication networks.

Moreover, by allowing Java applications to access resources within the network, the JAIN idea is shifting the communications market from many proprietary closed systems to a single network architecture where services can be rapidly created and deployed.

The JAIN service logic execution environment (JSLEE) [14, 15] is an integral part of the set of JAINs. It is the logic and execution environment in which communication applications are deployed to use the different network resources defined by the other JAINs. Basically, the JSLEE specification defines interfaces and requirements for communication applications relying on JAIN standards.

JSLEE

JSLEE is a standard architecture defining an environment targeted at communication-based applications.

The specification includes a component model for structuring the application logic of communications applications as a set of object-oriented components, and for arranging these components into higher-level and more complicated services. The programming language used by application developers in JSLEE is Java.

The SLEE architecture also defines the contract between these components and the container that will host these components at run time. The SLEE specification supports the development of highly available and scalable distributed SLEE specification-compliant application servers, even if it does not suggest any particular implementation strategy. More important, applications may be written once and then deployed on any application environment that implements the SLEE specification. The system administrator of a JSLEE controls the life cycle (including deployment, un-deployment, and on-line upgrade) of a service.

The life-cycle management is achieved through the use of the standard management interfaces provided by a compliant JSLEE, typically reusing Java management extension (JMX) techniques [16]. A service includes meta-information that describes it—i.e., its name, vendor, and version—and any program code associated with it. The program code can include Java classes and service building blocks (SBBs).

The atomic element defined by JSLEE is the SBB. An SBB is a software component that sends and receives events and performs computations based on the receipt of events and its current state.

SBBs are stateful components because they can remember the results of previous computations and those results can be applied in additional computations. SBBs perform logic that is based on the receipt of events. Events are used to represent occurrences of importance that may occur at arbitrary points in time. For example, the act of an external system delegating to the SLEE a call setup may occur at any point in time and is therefore easily modeled as an event.

An SBB definition includes meta-information that describes it (e.g. its name, vendor, and version), the list of events that it can receive, and Java classes that provide the logic of the SBB itself.

An event represents an occurrence that may require application processing. It contains information that describes the occurrence, such as the source of the event. An event may asynchronously originate from a number of sources—for example, an external resource such as a communications protocol stack, from the SLEE itself, or from application components within the SLEE.

Resources are external entities that interact with other systems outside of the SLEE, such as network elements (i.e., messaging server or SIP server). A resource adaptor adapts the particular interfaces and requirements of a resource into the interfaces and requirements of the JSLEE.

StarSLEE [7] is a prototype event-based execution engine for telecommunication applications inspired from JSLEE specification that reuse the concept of SBBs and resource adaptors; in this work, we applied dynamic AOP techniques for managing run-time monitoring on StarSLEE.

Aspect-Oriented Programming

Aspect-oriented programming (AOP) [5] is a new programming paradigm extending object-oriented software development. The main purpose of AOP is separation of concerns, developed orthogonally from the main functionality of a software system.

While the term "concern" represents whichever specific requirement to be implemented in a software system, crosscutting concerns are requirements whose implementation is difficult to modularize—e.g. security, persistence, logging, etc.—because the code involved by these concerns is scattered throughout several classes in an object-oriented application (see *Figure 1*).

Instead, with AOP, developers can remove scattered code related to crosscutting concerns from classes and place them

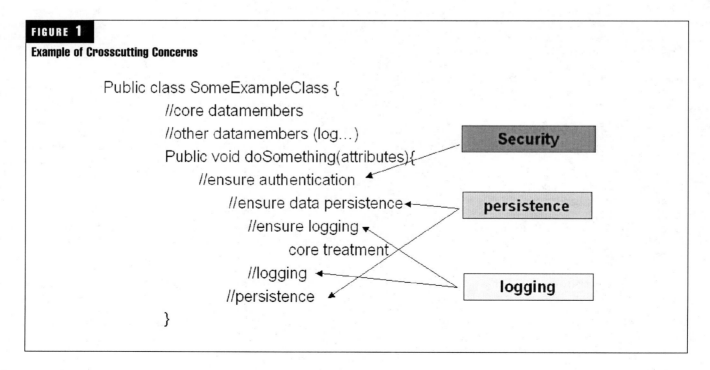

FIGURE 1

Example of Crosscutting Concerns

into first-class elements called aspects. This way, the original classes are no more responsible for managing functionalities not related to their core functionality. A direct consequence of aspect use is that less code needs to be written, so code that would otherwise be spread throughout the system can now be modularized in one place.

In *Figure 2*, it is easy to see that now the doSomething method contains only business-related code.

It means that now we are able to completely separate cross-cutting concerns from business ones. Thus, by keeping aspects separate from the target application methods they interact with, the application source code is easier to understand at the implementation level.

Therefore, with this new structure, if it is necessary to modify the logging-related code, this only needs to be changed in one place and not in each class.

AOP methodology is implemented by platforms such as AspectJ [10], AspectWerkz [11], AspectC++ [8], and JBoss–AOP [12], which are among the most stable and widespread AOP frameworks. All these tools rely on their own join-point model, which defines the points along the execution of a program that can possibly be addressed by an aspect.

Thus, AOP involves a compiling process called weaving for the actual insertion of aspect code into pre-existing application source code or byte code. Weaving can occur at compile time, load time, and run time.

In AOP terminology, an aspect is composed by a set of pointcuts and advices. The term "advice" represents the implementation of a crosscutting concern (i.e., additional code to be executed in particular points of the application code).

Advices can be of the following three types: before, after and around. A before advice is executed before the join point

(e.g., before method execution), an after advice is executed after the join point (e.g., after returning of method execution), and an around advice is executed instead of a join point (e.g., instead of the method body implementation).

AOP also involves means of identification of the join points to be affected by an aspect. The AOP term "pointcut" implicitly defines at which join points in the dynamic execution of the program extra code should be inserted. Pointcuts can describe sets of join points by specifying, for example, the objects and methods to be considered, or a specific method call or execution. Moreover, wild cards and logical operators can be used to combine pointcuts in more complex ones, identifying a wider set of join points.

The term "dynamic AOP" is attributed to platforms allowing the insertion and withdrawal of aspects at run time. This means that an aspect can be dynamically and remotely inserted and then further changed without stopping the application.

Moreover, AOP has been used to instrument source code and collect dynamic information about a system.

Putting together these features, in this work we have implemented a remote system monitoring and logging framework which is able to insert (and then change at run time) monitoring code in a JSLEE–distributed application.

The Monitoring Aspect in JBoss-AOP

JBoss-AOP is a Java framework for dynamic AOP that can be run inside or outside the JBoss Enterprise Application Server. For example, JBoss-AOP allows you intercepting a method call and transparently insert additional code (aspect) when the method is invoked.

All AOP constructs are defined as pure Java classes and bound to the application code via an XML [5] file containing the pointcut definitions.

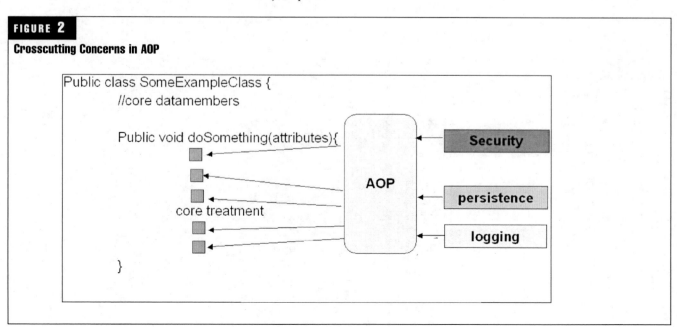

FIGURE 2

Crosscutting Concerns in AOP

FIGURE 3

An Example of Jboss-aop.xml File

```
<?xml version="1.0" encoding="UTF-8"?>
<aop>
<prepare
expr="execution(public void Foo->method())"/>
</aop>
```

This XML file (jboss-aop.xml) is read at process start-up by the JBoss container, which defines the maximum superset of join points that can be defined in application code, i.e., where an interception could occur at run time.

Figure 3 is an example of an XML file that prepares the body of method "method" of class "Foo."

After that, during application loading, JBoss prepares application classes, instrumenting their bytecode with the addition of "hooks," i.e., invocations to an aspect's advice.

If joint points are not instrumented in application code at first deployment, it will be impossible to bind them to a new interceptor at run time. Such a mechanism reduces overhead on the process at run time by limiting checks on joint points and enhances security by avoiding any code to be intercepted.

The JBoss-AOP framework is based on invocation objects implementing the Invocation interface.

Invocation objects are the representation of join points at run time. They contain run-time information about their join points and also drive the flow of aspects.

There are different invocation objects, including the following:

- MethodInvocation is created and used when a method is intercepted.
- ConstructorInvocation is created and used when a constructor is intercepted.

- FieldInvocation is an abstract base class that encapsulates field access.
- FieldReadInvocation extends FieldInvocation and is created when a field is read.
- FieldWriteInvocation extends FieldInvocation and is created when a field is written to.
- MethodCalledByMethod is allocated when using "call" pointcut expressions. This particular class encapsulates a method that is calling another method so that you can access the caller and callee.

Similarly, MethodCalledByConstructor and Constructor CalledByMethod are allocated respectively when a constructor is calling a method and vice versa.

In JBoss-AOP, an aspect is a class implementing the Interceptor interface. This class must implement two methods: getName, which returns the name of the aspect interceptor; and invoke, which represents the advice method and provides the invocation object as input.

The most important method of the Invocation interface is the invokeNext. Calling the invokeNext method means executing the intercepted method or constructor (or other) and returns the return value of that method if any.

Not calling that method will not execute the intercepted code, meaning overwriting it and interfering with the normal execution of the method.

As exemplified above, one can see multiple usages of interceptors by acting as before, after, or around advice based on when the invokeNext method is used.

Dynamic AOP hence becomes a powerful tool for application servers such as SLEE, enabling many monitoring applications such as testing, logging, service-level tracing, statistics gathering, bug fixing, etc.

Aspect Deployment on the Service Bus

The service bus [17] is an event-based distributed middleware that allows for run-time deployment and monitoring

---

FIGURE 4

An Example of Interceptor

```
public class InterceptorExample implements Interceptor {

    public String getName() {
        return "InterceptorExample";
    }

    public Object invoke(Invocation invocation) throws Throwable {
        System.out.println("Entering anything");
        return invocation.invokeNext(); // proceed to next advice or actual call
    }

}
```

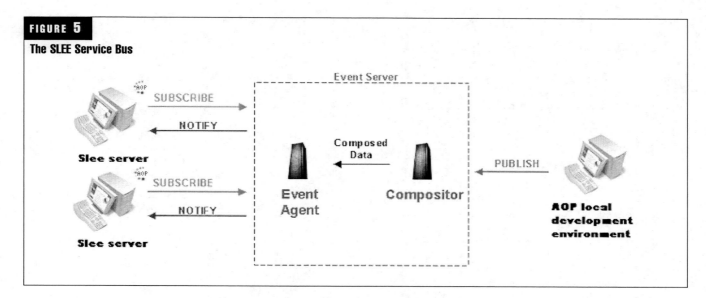

FIGURE 5

The SLEE Service Bus

of service-level information over heterogeneous resources of a communication network. In particular, it allows deploying aspects over several running SLEE containers at the same time, using a new type extension called Aspect. This overall mechanism enables the local development of aspects and their remote deployment onto the network.

The picture in the *Figure 5* describes the mechanisms used within the service bus for SLEE servers to subscribe to AOP–related information while an AOP deployment console publishes the command to deploy or undeploy aspects. The steps for deploying an aspect through the service bus are as follows:

FIGURE 6

Aspect Configuration in SLEE

```
<name>myAspect</name>
<pointcut>
execution(void myPackage.Aclass->method(..))
</pointcut>
<interceptor>test.InterceptorExample</interceptor>
<jar>http://anySite/interceptionExample.jar</jar>
```

- Write and compile the interceptor and put it into a JAR.
- Deploy the JAR file on to an HTTP server.
- Send a publish message through the service bus, indicating the target SLEE(s) and specifying the AOP-related information as follows: name—logical name, corresponding to a primary key, i.e. a unique identifier between an interceptor and a pointcut; pointcut—the pointcut used to intercept the classes to be monitored; interceptor name—the fully qualified name of the interceptor class; URL—the HTTP URL of the JAR file containing the interceptor class.

Figure 6 is an example of such AOP–related information published through the service bus to deploy an aspect on a target SLEE:

The following figures respectively display the successful installation (*Figure 7*) and the un-deploy command of an aspect on a SLEE (*Figure 8*) in the Eclipse-based service bus management console.

FIGURE 7

Monitoring Aspect Deployment

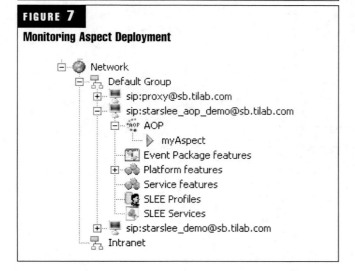

FIGURE 8

Monitoring Aspect Un-Deployment

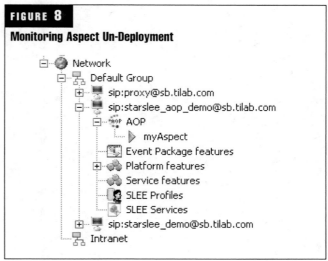

Discussion and Future Work

In JSLEE architecture, monitoring [13] is an indispensable technique to test the behavior of a system, debug the code, obtain usage statistics, or measure QoS.

Program instrumentation, which is typically a manual and time-consuming task, is often used to insert monitoring code into the system to be monitored before service deployment.

AOP has been already used to automate code instrumentation before deployment [4] but, with this approach, it is not possible to dynamically change monitoring code after service deployment.

In our approach, using dynamic AOP for managing monitoring tasks has revealed several advantages. A monitoring aspect is developed once and then deployed to different containers, using the Eclipse-based administrator console. The adaptation to different classes is eased by the power and flexibility of language-based constructs (pointcuts).

The added value of our approach is the use of a dynamic AOP framework. This allows the fast deployment of new monitoring aspects on different SLEE containers already running to dynamically modify their behavior.

Future enhancements of the monitoring platform will mainly target a friendly usage of this technology by creating a library of aspects of interest for SLEE containers, SBBs, and services, and the addition of wizards for handling aspect templates. Aspect templates parameters could be easily instantiated to ad-hoc aspects for the system to be monitored.

References

[1]   Licciardi, C.A., Falcarin, P., Analysis of NGN service creation technologies. In Annual Review of Communications vol. 57, IEC, December 2003.

[2]   Licciardi, C.A., Falcarin, P., Next Generation Networks: The services offering standpoint. In Comprehensive Report on IP services, Special Issue of the International Engineering Consortium, October 2002.

[3]   Glitho, R.H., Khendek, F., De Marco, A., Creating Value Added Services in Internet Telephony: An Overview and a Case Study on a High-Level Service Creation Environment. In IEEE Transactions on Systems, Man, and Cybernetics, Part C: Applications and Review, Vol. 33, n. 4, November 2003.

[4]   Mahrenholz, D., Spinczyk, O., and Schroeder-Preikschat, W. Program Instrumentation for Debugging and Monitoring with AspectC++. In Proceedings of the 5th IEEE International Symposium on Object-oriented Real-time Distributed Computing, Washington, D.C., USA, April 2002.

[5]   G. Kiczales, J. Lamping, A. Mendhekar, C. Maeda, C. Lopes, J. M. Loingtier, and J. Irwin, "Aspect-oriented programming," Proc. of 11th European Conference Object-Oriented Programming, 1997, pp. 220-242.

[6]   J. Rosenberg, H. Schulzrinne, G. Camarillo, A. Johnston, J. Peterson, R. Sparks, M. Handley, E. Schooler, SIP: Session Initiation Protocol, RFC 3261, June 2002.

[7]   A. Baravaglio, C.A. Licciardi, C. Venezia. Web Service Applicability in Telecommunications Service Platforms. In Proc. of the International Conference on Next Generation Web Services Practices, Seoul, Korea, August 2005.

[8]   AspectC++ project. On-line at www.aspectc.org.

[9]   XML (Extensible Mark-up Language) specification. On-line at www.w3.org/XML.

[10] AspectJ project. On-line at eclipse.org/aspectj.

[11] AspectWerkz project. On-line at aspectwerkz.codehaus.org.

[12] JBoss AOP framework. On-line at www.jboss.org/developers/projects/jboss/aop.

[13] Mahrenholz, D.: Minimal invasive monitoring. In Proceedings of Fourth IEEE International Symposium on Object-Oriented Real-Time Distributed Computing, May 2001 pp. 251 – 258.

[14] JAIN SLEE API Specification, Java Specification Request (JSR) 22, 1999. On-line at www.jcp.org/jsr/detail/22.jsp.

[15] JSLEE (JSLEE) v1.1, Java Specification Request (JSR) 240, 2004. On-line at www.jcp.org/jsr/detail/240.jsp.

[16] Java Management Extensions (JMX), On-line at java.sun.com/products/JavaManagement.

[17] G. Valetto, L.W. Goix, G. Delaire. Towards Service Awareness and Autonomic Features in a SIP–enabled Network. In Proc. of the Workshop on Autonomic Communication (WAC2005), Athens, Greece, October 2005.

Infrastructure as Platform: Services-Oriented Architecture, Virtualization, and 21st-Century Communications

Joe Mambretti

Director, International Center for Advanced Internet Research
Northwestern University

Abstract

New forms of services-oriented architecture enable advanced capabilities to be created by abstracting functionality from specific physical implementations, system configurations, and middleware. This ongoing macro trend toward services-oriented architectures is beginning to transform communications services at all levels. As convergence toward common protocols and digital communications infrastructure rapidly progresses, this new architecture offers another potential for rapid migration from legacy models. This type of architecture paves the way for new communication services and infrastructures that fundamentally depart from traditional approaches, which consist of highly defined capabilities closely coupled with supporting infrastructure. In contrast, new communications infrastructure can be designed as a large-scale distributed facility that can be used as a foundation platform, on which it is possible to create many types of networks, services, and applications that are not closely integrated with underlying physical components and system configurations. The separation of service layer functionality from the attributes of underlying infrastructure has far-reaching implications. For example, it allows capabilities for highly programmable communication services and networks. Core network elements can be used as separable components, dynamically mixed and integrated with a flexibility not possible using traditional infrastructure. Consequently, these components become flexible tool sets, middleware modules, and other resources that can be continually and dynamically assembled and reassembled to create and enhance services. This new architectural technique is beginning to emerge from research labs, and it is being demonstrated in prototype on metropolitan, national, and international facilities.

Introduction

In a very short time, information technology has undergone many revolutions, driven by multiple macros force—innovative disruptive architectures and technologies, large-scale gains in component performance, increasing cost efficiencies, and powerful new software functionality. These changes continually create new opportunities for applications and services creation. Historically, one important macro trend has been the creation of enhanced capabilities through architecture that provides for various levels of functional abstraction by making those capabilities less dependent on specific physical-layer implementations and configurations, including through the virtualization of infrastructure environments. For example, during the infancy of the computer revolution, programming was accomplished by directly manipulating physical elements. Later, increasingly more sophisticated compilers and programming languages were created, allowing continually more abstract software architectures, programmability, and general capability. Another example is the trend toward using new abstraction layers to separate service and application interfaces from authentication and authorization processes, as well as from specific implementations. This approach allows for significantly enhanced services and applications by avoiding closely integrating capabilities with specific systems, physical hardware, and configurations.

This general information technology trend continues today, and it can be observed in many of the changes influencing new communications architecture. In part, this trend is being propelled by convergence. For many decades, different communication modalities—e.g., voice, video, and data—were designed and implemented as distinctive services on separate, incompatible infrastructures, from end-points to edge equipment through core facilities. Today, all communications modalities are being migrated to a common digital infrastructure, supported by a ubiquitous set of protocols—e.g., transmission control protocol/Internet protocol (TCP/IP). This trend toward digital communications enables an ever-increasing level of abstraction between physical infrastructure and communication services. At the edge of the network, it is no longer necessary to have devices designed for a specific communication service modality, one for video, another for data, and another for voice communications. Any device should be able to support any service modality at any location. Recognition of this potential is driving a revolution in the next generation of communications-enabled consumer electronics.

However, many of today's designs for new digital services are still based on long-held assumptions about creating and deploying communication services. Traditionally, a limited set of these services is precisely designed, with requirements carefully measured and specified. Similarly, the infrastructure resources being designed to support these services are created to precisely match those service requirements. An expectation is that this infrastructure can be deployed and supported for many years without major changes. Ultimately, this approach may result in an infrastructure that has many elements of the inherent inflexibility of the traditional infrastructure that it is replacing. Already, applications are being designed that cannot be supported by today's Internet, even by many of its more advanced implementations.

A different approach can leverage the many opportunities provided by digital communications convergence—opportunities to fundamentally change the traditional approach to the design and implementation of services. A new architectural framework can be created that enhances abstraction levels and reduces dependencies on specific communication infrastructure implementations such as hardware, protocol, and systems. By creating high degrees of separation between end-delivered services and support infrastructure, communication services can be provided with substantially more flexibility and capabilities. Because of this higher level of abstraction, core infrastructure becomes a programmable facility, or platform, on which it is possible to create many new types of functions. This architecture can be used to create truly programmable large-scale services, applications, and networks.

Application Drivers

As with many prior innovative trends, development of this new architecture is being driven in large part by multiple large-scale, resource-intensive applications—especially global science applications—that cannot be supported by traditional infrastructure [1]. Many require significant communication resources, such as asymmetric bandwidth intensive data transfers among sites around the world. Other applications are highly intolerant of latency and require high-performance deterministic end-to-end channels. Also, some applications require continually changing real-time infrastructure. In addition, many new types of general enterprise and consumer communication services require capabilities that are not available through traditional archi-

tecture. For example, current IP–based digital media services tend to be implemented as distinct applications. A more flexible infrastructure can allow multiple digital media capabilities to be easily integrated into virtually any other service or individual application and integrated into global communication services. For example, this approach would enable common portals to easily implement interactive digital media. This architecture will also enable digital media to stream with significantly higher density than common modalities used today.

A New Architectural Model

The advanced networking research community is exploring a fundamentally new approach to the design of communications infrastructure. Rather than design a network infrastructure specifically for a limited set of communication services, these new initiatives are conceptualizing basic infrastructure as a large-scale distributed facility that can be used as a platform for creating, programming, and reprogramming limitless new specialized networks and distinctive services, including those not yet invented. One general goal is enabling any service to be accessed by any device at any location. The key to this potential is a new architecture that abstracts service functionalities from the inflexibilities and restrictions of supporting physical infrastructure, in part by allowing for levels of virtualization not previously possible. A high level of abstraction can be provided by resource layers that are transparent to higher-layer services, for example, by using virtualization techniques. Consequently, communication services can be designed independently from specific core, access, and edge delivery facilities and from the specific characteristics of edge devices.

The separation of services from underlying infrastructure has multiple implications. This architecture is significantly more flexible and scalable than traditional designs and allows for an infrastructure that is unlimited in its service creation capabilities. Instead of creating services on a centralized carrier network, services are based on a large-scale, distributed facility that advertises, allocates, manages, and controls resources dynamically.

Using this new architecture, new services can be more easily created and reprogrammed by continually assembling and reassembling multiple communication resource elements as required. This architecture also provides a potential for many networks to co-exist in an infrastructure, and it provides opportunities for a high level of services specialization, differentiation, and customization. In addition, it provides a means to separate physical infrastructure provisioning, management, operations, and support from end-delivered services. Consequently, it can allow some organizations to focus completely on infrastructure provisioning while allowing their customer organizations to provide only end-delivered services without having to own or operate their own infrastructure.

Until recently, the architecture and technology required to reduce the dependencies of communication services on supporting infrastructure did not exist. However, the current trend toward protocol and digital infrastructure convergence has provided major capabilities for mitigating or even eliminating these dependencies. In addition, other opportunities for dependency reduction are being created through new protocols, methods, and technologies being designed for every level of network interface, access paths, and core infrastructure.

Interface Drivers: Services-Oriented Architecture and Web Services Resource Framework

Considerable standards and development activities are focused on services-oriented architecture, which provides a common method of defining and implementing capabilities within information technology (IT) environments. Any resource in such an environment can be an advertised "service." Web services architecture provides methods for defining "packages" of such services, that is, mechanisms that can be used to gather and use multiple individual services. These and related architectures and methods are powerful tools for creating capabilities that are abstracted from the specifics of IT environments and are especially useful for communications services.

Multiple standards bodies are addressing common architectures, methods, and information sharing for Web services, a Web services data language (WSDL), and related architecture. For example, the World Wide Web Consortium (W3C) is developing concepts of a "semantic Web," which provides a way of enabling Web information to be easily understood by system processes. This consortium defines the "building blocks" required for common interoperability methods such as extensible markup language (XML) and related standard message exchange mechanisms such as simple object access protocol (SOAP).

The Organization for the Advancement of Structured Information Standards (OASIS) is developing a Web service resource framework (WSRF) standard. WSRF provides Web service specifications that define a method of modeling and managing state within a Web service context. WSDL schema can provide definitions of packaged functionality, including stateful resource and service elements, and edge processes. The Global Grid Forum—the standards body for grid technology, which advances capabilities for flexible distributed environments—has adopted the WSRF architecture as a convenient top-level abstraction method, which is now part of their open grid services architecture model [2].

These techniques can be used to enable edge processes to directly provision and manage network services and other resources. Using XML for schemas for common service definitions provides a powerful abstraction method for communication services. It allows for common capabilities to be created and provisioned across multiple heterogeneous domains. The International Telecommunication Union Telecommunication Standardization Sector (ITU–T) ASN.1 committee is developing a new standard that uses the ASN.1 language (a traditional notation for defining protocol messages that is commonly used in the communications industry) as a new schema definition language for XML (www.itu.int).

ASN.1 provides for a clear distinction between content—e.g., "abstract syntax" ("message description," which contains no implied coding method) and "transfer syntax" ("encoding") (http://asn1.elibel.tm.fr/en). However, using these techniques, communication service designers do not have to adhere to common definitions. They can base services on commonly defined elements, create new elements, or use mixtures of common and customized elements. They can be integrated with signaling used for general or core provisioning methods, e.g., as defined by the ITU–T's automatically switched transport network (ASTN) (G.807) and automatically switched optical network (ASON) (Y.1304) architectural standards (www.itu.int).

Services Signaling

One major trend enhancing capabilities for abstracting communication services from specific infrastructure implementations is the reduction of hierarchical communication protocol layers, even to the point of directly placing data on light paths—for example, through IP–over–DWDM services. A related trend also enhancing abstraction opportunities is the migration of signaling architecture—for both in-band and out-of-band communications—to IP–based standards, which significantly enhances the potential for abstracting communications services from specific physical implementations. Signaling based on IP can be used for access control, topology discovery, traffic engineering, wavelength routing, reconfiguration, protection and restoration, and for the configuration and re-configuration of many individual infrastructure components. The functional abstraction enabled by these trends allow for substantially more flexibility in services creation and deployment because such services no longer have to be dependent on a centralized governance model.

Traditionally, the creation and deployment of new communication services has been dependent on centralized management and control capabilities. Because of this dependency, the creation and deployment of new services has been a slow, multi-year process. New architectural models that abstract services from infrastructure enable core resources to be used as a platform on which it is possible to create many new types of services and applications independent of centralized management processes. These communication infrastructure platforms provide for various advertised capabilities, such as application programming interfaces (APIs), which can be used to create new services. These types of capabilities are being created in early prototypes [3].

For example, the experimental optical dynamic intelligent network (ODIN) services architecture was designed, developed, and implemented on an optical test bed to demonstrate the potential for a service layer that could broker resource requests by applications and services for core network services, primarily dynamically allocated dedicated Layer-2 (L2) circuits and L1 light paths [4]. A series of experiments and demonstrations have shown that data-intensive applications can use these signaling and service-layer techniques to provision light paths dynamically as required on optical networking accessible through specialized APIs. In these demonstrations, ODIN served as an intermediary service layer between the applications and lower-level network components. Essentially, ODIN extends control-plane functionality through the service layer to the application, including functions such as light-path addressing, dynamic path computation, resource discovery and light-path reachability, as well as other functions [5]. The actually optical level provisioning tool used is the Internet Engineering Task Force (IETF) generalized multiprotocol label switching (GMPLS) standard. The signaling protocol used in part is based on the experimental lightweight path control (LPC) protocol, described in an IETF draft (www.ietf.org) [6]. The LPC protocol provides a standard mechanism that allows edge processes to communicate requirements for specific network paths, including light paths, through a server-based process that directly establishing the paths using user network interfaces (UNIs). Request signals are interpreted by a server that has direct access to network state information and that can establish topologies dynamically.

The new potentials for abstracting communication services from physical infrastructure are being made possible by the next generation of optical components, which are more flexible than traditional devices. Dense wavelength division multiplexing (DWDM) has supported static optical channels for two decades. However, new components, protocols, and software provide capabilities for dynamic provisioning. Emerging systems are being introduced that include multifunctional optical cross-connects (OXCs)—e.g., those that are optical-to-electrical-to-optical (OEO)–based and those that are optical-to-optical-to-optical (OOO)–2D– and 3D–micro-electromechanical system (MEMS)–based, addressable DWDM interfaces, controllable optical add/drop multiplexers (OADMs), tunable lasers, tunable amplifiers, flexible gatings, and other devices.

Experiments on the potential for abstracting services from physical implementation using such new components are being conducted on advanced optical test beds. One such test bed is the Optical Metro Network Initiative (OMNI) test bed, a wide-area metro photonic test bed in the Chicago area that supports 24 10 GE optical channels among four core node sites interconnected with dedicated fiber (www.icair.org/omninet). These nodes are comprised of a DWDM photonic switch (2D–MEMs–based), an optical fiber amplifier (OFA, to compensate for link-and-switch decibel loss), optical transponders/receivers (OTRs), and high-performance L2/L3 routers/switches. Each node supports 4x10 Gbps optical channels, based on addressable wavelengths. This test bed is being used to demonstrate the utility of creating communication services using a highly flexible core infrastructure.

OptIPuter

These concepts are also being explored on larger-scale test beds, such as the national-scale OptIPuter, an experimental research initiative that is being funded by the National Science Foundation [7]. In this project, the reference requirements for communication services are those that are needed by large-scale science projects such as geophysical sciences, bioinformatics, and space exploration. In part, the OptIPuter has been designed as a next-generation supercomputer. Traditionally, supercomputers have been designed and developed as a specific environment on a defined physical infrastructure that can support large-scale, computation-intensive applications. For this project, the high-performance computing and communications "platform" consists of a national-scale-distributed environment,

based on a national optical network, designed with an innovative architecture that closely integrates light paths, IP signaling and data communication services, mass data-storage systems, high-performance cluster-based computational processing, and advanced visualization technologies. Although these resource components are integrated, the architecture within this environment provides for a high level of services abstractions.

The OptIPuter was designed as a large-scale, high-performance, distributed environment, a "lambda grid," within which it is possible to create many types of supercomputers. Traditionally, computation-intensive applications must conform to restrictions inherent in physical attributes of supporting infrastructure. The design of the OptIPuter environment allows core services, through specialized middleware, to be abstracted so that applications can create ad hoc virtual supercomputers designed to meet their specific requirements [8]. The central architectural resource and key enabling component for this environment consists of multiple optical networks, not computers, based on dynamic light-path provisioning. These light paths enable the dynamic instantiation of "supernetworks," which function essentially as distributed backplanes for virtual environments on nationwide or worldwide fabrics.

User-Controlled Light-Path Architecture

For many years, the Canadian Network for the Advancement of Research Industry and Education (CANARIE) has been designing and developing innovative communication capabilities based on service-layer abstractions. They are recognized as world leaders in bringing these concepts from research labs into production services, e.g., on the CA*net 4 network (www.canarie.ca/canet4). For example, the CANARIE user-controlled light path (UCLP) architecture was created to allow a service provide to allocate a segmented network domain within a larger distributed facility to end customers. UCLP enables edge processes to discover, access, provision, and dynamically reconfigure optical (L1) light paths within a domain or across multiple domains, independent of any central authority.

The UCLP design, process, and software provides a mechanism not just for allocating facility resources, e.g., light paths, but also the control, management, and engineering systems for those resources. In addition, this architecture provides capabilities that can enable customers to further reallocate subsections of those resources along with their related control, management, and engineering systems. The UCLP design was created within the context of services-oriented architecture.

Global Lambda Integrated Facility

These architectural concepts are also being explored on an international scale. An international consortium is designing and developing the Global Lambda Integrated Facility (GLIF) as an international platform that can provide for a set of core basic resources within which it is possible to create multiple differentiated specialized networks and services. The GLIF provides for a closely integrated environment (networking infrastructure, network engineering, system integration, middleware, applications, etc.) and support capabilities based on dynamic configuration and reconfiguration of resources. GLIF is based on a fabric consisting, in part, of dynamically allocated light paths, some of which are individually addressable wavelengths. These light paths are globally interconnected through sites worldwide that provide advanced services, including high-performance transport, based on an open optical exchange point architecture.

For example, the StarLight facility in Chicago has been designed, developed, and implemented to provide next-generation communication services based on new services abstraction architecture (www.startap.net/starlight) [9]. Similar facilities are also being developed in other cities, such as NetherLight in Amsterdam, Netherlands, UKLight in London, England, NorththenLight in Stockholm, Sweden, and CzechLight in Prague, Czech Republic [10]. Related exchanges have also been established in Japan, Korea, China, and several countries on other continents. This architecture is being developed so that multiple edge processes can access and manage a wide range of specialized services, e.g., through middleware integrated with control and management planes, including multiple based on core resources, such as high-performance computational clusters, rendering and visualization facilities, mass storage systems, instrumentation, and L1 paths on a dynamic, reconfigurable optical infrastructure [11]. Using these architecture options can be provided for either accepting default pre-defined services or for creating customized new services within distributed environments that provide component resources that can be dynamically discovered, gathered and integrated using specialized access techniques, such as Web services based signaling based on semantic web architecture.

iGRID2005

As noted, many of the architectural concepts described here are being driven by global high-performance applications. Recently, iGRID2005 (igrid2005.org) presented an international showcase of 49 next-generation applications from 20 countries requiring high-performance communications provisioned on a specialized, flexible global infrastructure. The majority of the applications demonstrated could not be supported by a traditional data communications infrastructure. For many of the applications demonstrated, key enabling factors consisted of the enhanced flexibility provided by communication services abstractions, including those that allowed access to deterministic paths provisioned on a global 200 Gbps infrastructure customized specifically for the event and provided primarily by GLIF resources. Although the provisioning for the event as a whole was not automated, a number of individual application demonstrations employed new types of provisioning based on advanced architecture for service abstractions interlinked with core network resource provision, including light-path provisioning.

This event demonstrated the innovative applications that are possible when dependencies between services or applications and particular communications infrastructure are mitigated or removed. For example, many applications showcased at the event demonstrated the value of service virtualization in communication environments. In the future, the design of new applications and services will benefit from capabilities that provide increasingly higher levels of programmability and virtualization.

Summary

The creation of enhanced architectural abstractions for information technology resources has been a major force shaping IT development for decades. Today, this approach is influencing the direction of new communications architecture. Traditionally, communication providers have defined and provided services that have been closely coupled with supporting physical infrastructure. A problem with this traditional model is that new services creation and enhancement is slow, costly, and complex. By creating an architecture that allows for a significant level abstraction to separate services from infrastructure, the design and creation of services can be much more independent of existing physical implementation and configurations. The physical infrastructure can become a large-scale distributed facility that, in effect, becomes a programmable platform for many new types of communication services. This architecture can also lead to new types of business models—for example, some providers may only offer basic infrastructure and allow other organizations to offer higher-layer services. They may also provide a wide range of accessible APIs so that multiple organizations, and even individuals, can create customized services. This approach is already emerging within various commercial Web-based services—for example, among general-content providers that are allowing anyone on the Internet to access resource APIs. In the future, this type of architecture will enable higher-level abstractions to provide new types of innovative services by integrating these higher-level capabilities with those that can directly and dynamically reconfiguring basic communications infrastructure resources.

References

[1] Special Issue, Journal of Future Generation Computer Systems, Elsevier Science Press, Vol. 19, Issue 6, Aug 2003.

[2] I. Foster, C. Kesselman, "The Grid: Blueprint for New Computing Infrastructure," Morgan Kaufmann, 2003; I. Foster, C. Kesselman, J. Nick, and S. Tuecke, "The Physiology of the Grid: An Open Grid Services Architecture for Distributed Systems Integration," Open Grid Service Infrastructure WG, Global Grid Forum, June 2002. (www.globus.org/alliance/publications/papers/ogsa.pdf).

[3] J. Mambretti, J. Weinberger, J. Chen, E. Bacon, F. Yeh, D. Lillethun, B. Grossman, Y. Gu, M. Mazzuco, "The Photonic TeraStream: Enabling Next Generation Applications Through Intelligent Optical Networking at iGrid 2002," Journal of Future Computer Systems, Elsevier Press, August 2003, pp.897–908.

[4] J. Mambretti, "Experimental Optical Grid Networks: Integrating High Performance Infrastructure and Advanced Photonic Technology with Distributed Control Planes," Proceedings, Optical Networking for Grids Workshop, European Conference on Optical Communications, Stockholm, Sept. 4, 2004.

[5] D. Lillethun, J. Lange, J. Weinberger, Simple Path Control Protocol Specification, www.ietf.org/internet-drafts/draft-lillethun-spc-protocol-01.txt.

[6] L. Smarr, A. Chien, T. DeFanti, J. Leigh, P. Papadopoulos, "The OptIPuter," Special Issue: Blueprint for the Future of High Performance Networking Communications of the ACM, Vol. 46, No. 11, Nov. 2003, pp. 58–67.

[7] DeFanti, T., Brown, M., Leigh, J., Yu, O., He, E., Mambretti, J., Lillethun, D., and Weinberger, J., "Optical Switching Middleware for the OptIPuter," Special Issue on Photonic IP Network Technologies for Next-Generation Broadband Access. IEICE Transactions on Communications, E86-B, 8 (Aug. 2003), pp. 2263–2272.

[8] T. DeFanti, C. De Laat, J. Mambretti, Bill St. Arnaud, "TransLight: A Global Scale Lambda Grid for E-Science," Special Issue on "Blueprint for the Future of High Performance Networking," Communications of the ACM, Nov. 2003, Vol. 46, No. 11, pp. 34–41.

[9]   J. Mambretti, "Progress on TransLight and OptIPuter and Future Trends Towards Lambda Grids," Proceedings, Eighth International Symposium on Contemporary Photonics Technology (CPT2005)," Tokyo, Jan. 12–14, 2005.

[10]   J. Mambretti, "Ultra Performance Dynamic Optical Networks and Control Planes for Next Generation Applications," Proceedings, Mini-Symposium on Optical Data Networking, Grasmere, England, Aug. 22–24, 2005.

Executive Perspectives

Ubiquitous Broadband Access Networks with Peer-to-Peer Application Support

Benny Bing

Associate Director
School of ECE, Georgia Institute of Technology

Abstract

Many last-mile access technologies exist, including cable, DSL (digital subscriber line), powerline (the world's largest infrastructure), free-space optics, and broadcast satellite (which is attractive for both urban and rural areas, even isolated areas not serviceable by terrestrial methods). Unfortunately, these technologies represent a significant bottleneck between a high-speed residential network and a largely overbuilt core backbone network. This in turn makes it difficult to support end-to-end quality of service (QoS) for a wide variety of applications, including emerging peer-to-peer (PTP) applications and non-elastic applications such as voice and video, which cannot tolerate variable or excessive delay or data loss. This paper presents an architecture to help realize widespread broadband access connectivity with PTP application support. The importance of combining multihop wireless access with multi-wavelength optical access will be illustrated.

Introduction

When it comes to last-mile access, network operators have a difficult choice among competing technologies—DSL, cable, optical, and wireless. Key considerations to the choice include deployment cost and time, service range, and performance. The most widely deployed solutions today are DSL and cable modem networks. Although they offer better performance over 56 kbps dial-up telephone lines, they are not true broadband solutions for several reasons. For instance, they may not be able to provide enough bandwidth for emerging services such as PTP applications, multi-player games with audio/video chat to teammates, streaming content, high-definition video on demand, and interactive TV. In addition, fast Web-page download still poses a significant challenge, particularly when multimedia information is involved. Finally, only a handful of users can access multimedia files simultaneously, which is in stark contrast to direct broadcast TV services. To encourage broad

use, a true broadband solution must be scalable to thousands of users and must have the ability to create an ultra-fast Web-page download effect, superior to turning the pages of a book or flipping program channels on a TV, regardless of the content.

Broadband Optical Access

DSL or cable-modem access provides benefits of installed infrastructure, virtually eliminating deployment costs. If fixed wireless access is chosen, network providers gain the benefit of quick and flexible deployment. However, these access methods may suffer bottlenecks in bandwidth-on-demand performance and service range. For example, cable networks are susceptible to ingress noise, DSL systems can be plagued with significant crosstalk, and unprotected broadcast wireless links are prone to security breach and interference. Furthermore, current DSL and cable deployments tend to have a much higher transmission rate on the downstream link, which restricts Internet applications to mostly Web browsing and file downloads.

While wireless access is excellent for bandwidth scalability in terms of the number of users, optical access is excellent for bandwidth provisioning per user. Furthermore, the longer reach offered by optical access potentially leads to more subscribers.

Optical access networks offer symmetrical data transmission on both the upstream and downstream links, allowing subscribers to provide Internet services such as file sharing and Web hosting. The deployment costs still require laying fiber, which makes optical access networks more expensive to install. However, since fiber is loss-limited rather than bandwidth-limited (as opposed to copper wire, cable, and wireless systems), the potential performance gains and long-term prospects make optical access networks suitable for enterprises and new neighborhoods or installations. In addition, there are innovative solutions for deploying fiber in the last mile, even in established neighborhoods. For example, instead of investing in expensive dedicated fiber conduits, existing sanitary sewers, storm drains, water lines, and natural-gas lines that reach the premises of many end users can be exploited [4].

To provide bandwidth that is scalable with the number of subscribers, it is important to identify architectures that allow low equipment cost per subscriber. As new applications appear and demand higher bandwidth, the network should be gracefully upgraded. It is important to be able to perform an incremental upgrade where only the subscribers requiring higher bandwidth are upgraded, not the entire network. Since the lifespan of optical fiber plant is longer compared to copper or coaxial cables, it is expected that the optical network will be upgraded multiple times during its lifetime. As such, it is important to design access architectures that allow seamless upgrade. In addition, the deployment of fibers between residences can be used to connect subscribers directly [1], forming an autonomous network among residential end users, thereby improving service reliability while localizing PTP traffic traveling within the neighborhood. This paper, however, proposes a more cost-effective alternative in using multihop wireless to interconnect subscribers (see Integrated Broadband Access Architecture).

Passive Optical Networks

Passive optical networks (PONs) are viewed by many as an attractive solution to the last mile-problem, providing triple-play (data, voice, and video) services to end users at bandwidths far exceeding current access technologies. Unlike other access networks, PONs are point-to-multipoint networks capable of transmitting more than 20 kilometers of single-mode fiber. PONs minimize the number of optical transceivers, central office terminations, and fiber deployment compared to point-to-point and curb-switched fiber solutions. By using passive components (such as optical splitters, couplers, and array waveguide gratings [AWGs]) and eliminating regenerators and active equipment normally used in fiber networks (e.g., curb switches and optical amplifiers), PONs reduce the installation and maintenance costs of fiber as well as connector termination space. The general PON architecture consists of the optical line terminator (OLT) on the service provider side and optical network unit (ONU) (or the optical network terminal [ONT]) on the subscriber side. The ONU is connected to the OLT through one shared fiber and can take different fiber-to-the-x (FTTx) configurations—e.g., fiber-to-the-home (FTTH), fiber-to-the-curb (FTTC), and, more recently, fiber-to-the-premises (FTTP).

PONs typically fall under two groups: asynchronous PONs (APONs) and Ethernet PONs (EPONs). APONs are supported by full-service access networks (FSANs) and the International Telecommunication Union–Telecommunication Standardization Sector (ITU-T) because of its connection-oriented QoS feature and extensive legacy deployment in backbone networks. EPON is standardized by the Institute of Electrical and Electronics Engineers (IEEE) 802.3ah Ethernet in the First Mile (EFM) Task Force. EPONs leverage on inexpensive, high-performance, silicon-based optical Ethernet transceivers. Ethernet is gaining popularity and wide industry support because of the ubiquity of Ethernet deployment,

including gigabit Ethernet LANs, metro Ethernet, and wireless Ethernet such as Wi-Fi. EPONs ensure that Internet protocol (IP)/Ethernet packets start and terminate as IP/Ethernet packets without expensive and time-consuming protocol conversion or tedious connection setup.

Wavelength Division Multiplexing Optical Access

Wavelength division multiplexing (WDM) is a high-capacity and efficient optical signal transmission technology that is prevalent in long-haul backbone applications but is now emerging in metropolitan-area networks (MANs). WDM uses multiple wavelengths of light, each wavelength corresponding to a distinct optical channel (also known as lightpath or lambda,l), to transmit information over a single fiber-optic cable. Current WDM systems are limited to between 8 and 40 wavelengths on a single fiber. It is an economical alternative to installing more fibers and a means to dramatically improve data rates.

WDM optical access is a future-proof last-mile technology with enough flexibility to support new, unforeseen applications. WDM switching can dynamically offer each end user a unique optical wavelength for data transmission as well as the possibility of wavelength reuse and aggregation, thereby ensuring scalability in bandwidth assignment. For instance, heavy users (e.g., corporate users) may be assigned a single wavelength, whereas light users (e.g., residential users) may share a single wavelength, all on a single trunk fiber (*Figure 1*).

A WDM Broadband Optical Access Protocol

The benefits of PONs can be combined with WDM, giving rise to WDM PONs that provide increased bandwidth and allow scalability in bandwidth assignment. Key metrics in the physical layer (PHY) performance of WDM PONs include latency, link budget, transmitter power and passband, receiver sensitivity, number of serviceable wavelengths, and distance reachable.

The media access control (MAC) protocol helps resolve access contentions among the upstream transmissions from subscribers and essentially transforms a shared upstream channel into a point-to-point channel. Designing the MAC layer for a WDM PON is most challenging since the upstream and downstream channels are separated, multiple wavelengths can be used, and distances can go up to 20 km. The basic requirements are a method to maintain synchronization, support for fixed and variable-length packets, and the ability to scale to a wide range of data rates and services. The key to designing efficient MAC protocols is the identification of subscribers with data to transmit since these are the users who will use bandwidth.

Most PON protocols either use reservation protocols (where minislots are used to reserve larger data slots) or static time division multiple access (TDMA) protocols (where each ONU is statically assigned a fixed number of data slots). The access scheme presented here combines the advantages of reservation and static TDMA by preallocating a minimum number of data slots for the ONU, which can be increased dynamically in subsequent time division multiplex (TDM) frames through reservation minislots on a needed basis. By using reservation, we ensure that bandwidth can be varied dynamically. By preallocating data slots, a data packet can be transmitted immediately instead of waiting for a duration equivalent to the two-way propagation delay associated with the request and grant mechanism in reservation protocols. This is critical for supporting real-time PTP, voice, and video streaming. However, the number of data slots that can be preallocated must be kept to a minimum, because if more slots are preallocated, then there is a possibility that slots could become wasted when the network load is low and the traffic characteristics are bursty. This is the main disadvantage of static TDMA. In [7], a simple compression scheme is employed to minimize the number of preallocated slots.

To illustrate the delay improvement associated with bandwidth preallocation, a simulation experiment is conducted with both ATM and Ethernet traffic types at each ONU. In the case of ATM traffic, an arrival corresponds to a single ATM cell, whereas for Ethernet, each variable-size Ethernet packet is modeled as a batch (bulk) arrival of ATM cells. For the single packet arrival case, arrival pattern is Poisson with the link rate chosen to be 155.52 Mbps shared by 10 ONUs. A data cell comprises three header bytes and fifty-three data bytes, and a TDM frame comprises fifty-three data slots (equivalent to the duration of fifty-three data cells of fifty-six bytes each) with one data slot allocated for minislot reservation (i.e., fifty-two slots allocated for data cells). This gives a TDM frame duration of 53 x 56 x 8 bits / 155.52 Mbps = 152.67 ms and a slot time of 152.67 ms / 53 = 2.8807 ms. Since the TDM frame duration is typically designed for the maximum round-trip (two-way propagation) delay, our TDM frame duration is equivalent to a max-

imum distance of $(152.67 \times 10^{-6} \text{ s}) \times (2 \times 108 \text{ m/s}) / 2$, or roughly 15 km. The traffic load refers to the ratio of the data generation rate over the link rate. The cell arrival interval in each ONU is exponential distributed with mean $10 \times (53 \times 8)/(\text{link rate} \times \text{load})$. For the batch packet arrival case, the TDM frame and network settings are the same as before. The only change is the traffic pattern where we have simulated Poisson arrivals at each ONU with batch size uniformly distributed between 1 and 30 data cells (30 data cells is roughly equivalent to one maximum-length Ethernet packet of 1,500 bytes). The batch arrival interval is exponential distributed with mean $10 \times (53 \times 8)/(\text{link rate} \times \text{load} \times \text{average batch size})$. For the preallocation case, each ONU is preassigned a minimum of one data slot. A request packet transmitted on a minislot increases this minimum number to a number indicated in the request packet. For the case without pre-allocation, all data slots are assigned only after a request is made on a minislot. For static TDMA, five cells are allocated for each ONU, with the remaining three slots of the TDM frame unoccupied.

The performance of the case when each ONU has a single packet arrival is shown in *Figure 2A*. Preallocating data slots clearly reduces the average delay, even when the network is heavily loaded. Under high load, the performance of both protocols converges, which is expected since all data slots in a TDM frame tend to become filled continuously. Static TDMA performs best under low load due to the single arrival assumption. As can be seen from *Figure 2B*, static TDMA performs poorly because the average batch size is 15.5 cells per packet, which implies an average of three TDM

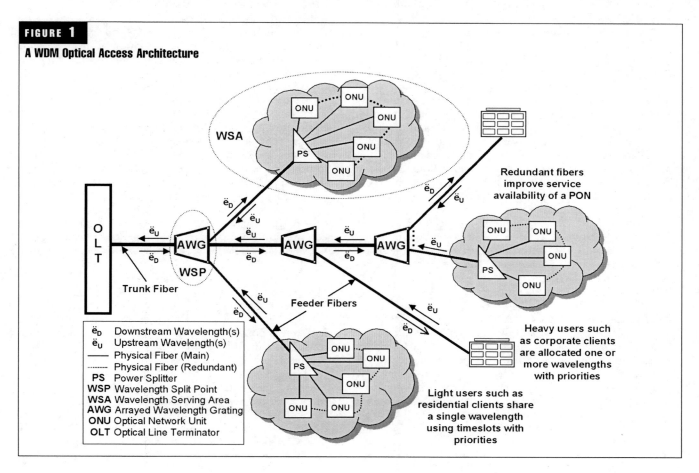

FIGURE 1

A WDM Optical Access Architecture

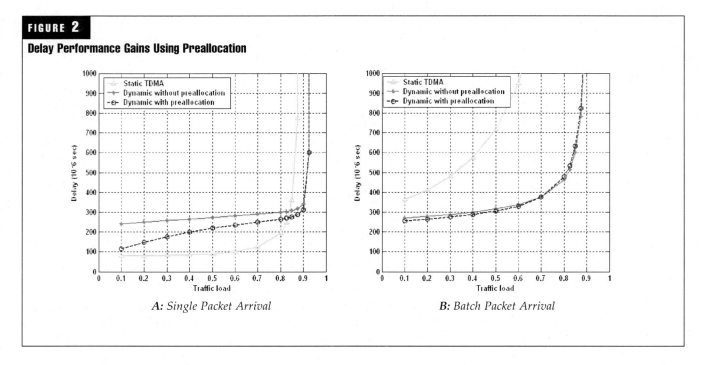

A: *Single Packet Arrival* **B:** *Batch Packet Arrival*

frames are needed to transmit all cells in a packet. Since each packet arrives randomly, in static TDMA, ONUs are not able to request for more bandwidth appropriate for the batch size.

Quality of Service Provisioning for Broadband Optical Access Networks

Unlike metro and long-haul networks, optical access networks must serve a more diverse and cost-sensitive customer base. End users may range from individual homes to corporate premises, hotels, and schools. Therefore, services must be provisioned accordingly. Data, voice, and video must be offered over the same high-speed connection with guarantees on QoS and the ability to upgrade bandwidth and purchase content on a needed basis. In addition, QoS provisioning in optical access networks must aim to match the vision of agile, high-capacity, and inexpensive metro optical networks against the practical operational reality of existing infrastructures deployed by telecom carriers. There is also a strong imperative for metro optical networks driven to extend broadband connectivity to end users. This can be accomplished by augmenting capacity, and, more important, introducing new technologies with strong price/service benefits that can support emerging broadband data services in a scalable and cost-effective manner, such as virtual private networks (VPN), voice over IP (VoIP), and virtual leased lines (VLLs).

Broadband Wireless Access

Choosing where to deploy fiber is a business and engineering decision today, and regulators are no longer involved in that decision. Choosing fiber for the enterprise and new neighborhoods are fairly straightforward decisions. However, deploying fiber in existing neighborhoods with existing sidewalks, sprinkler systems, and other infrastructure can get very expensive. Fiber splicing also increases network maintenance overheads such as costs and time. Wireless solutions are a good complementary technology.

Wireless solutions for the last mile have been labeled as "disruptive" technologies, since many phone companies are losing their landline business to wireless (just as they are losing business to VoIP). However, wireless solutions cannot compete with fiber solutions in terms of bandwidth provisioning, particularly as the operational range increases. They are also more prone to eavesdropping, security attacks, and traffic analysis, and are fundamentally limited by signal interference among concurrent transmissions [2]. On the other hand, the broadcast nature of the wireless medium offers ubiquity, user mobility, and immediate access. Besides offering a low-cost alternative to installing cable or leased telephone lines and allowing fixed-line operators to extend their broadband networks, wireless solutions also allow a long-distance carrier to bypass the local service provider, thereby cutting down subscriber costs. They are indispensable when wired interconnections are impractical because of rivers, rough terrain, private property, and highways.

When implemented effectively, wireless approaches can greatly simplify network management by reducing network configuration complexity; reducing dependence on gateways; and eliminating overheads associated with moves, additions, and changes.

Wireless Access Benefits

The use of wireless technology has the potential to significantly reduce the cost of broadband residential access. A substantial portion of the cost of last-mile deployment is in interconnecting the customer's premise to the central office. A typical cable TV (CATV) connection, for example, requires long cascades of about 30 trunk and line amplifiers between the head end and the customer premise, in addition to numerous passive taps. Reliability is a serious problem since all amplifiers were in series, and failure of any one device will result in a downstream signal outage from that point. A wireless solution not only removes the labor, material, and equipment costs associated with cabling, but also offers the flexibility to reconfigure or add more subscribers to the network without much planning effort and the cost of recabling, thereby making future expansion and growth inexpensive and easy. Moreover, the inherent flexibility of wireless communications to reconfigure quickly is an important consideration for increasing the reliability of network connectivity during emergencies and catastrophic events [1]. Another key advantage of wireless technology is the speed of deployment. In addition, a large portion of the deployment costs is incurred only when a customer signs up for the service [2].

There are emerging broadband wireless technologies that have realized impressive efficiencies of the order 10 bits/s/Hz, and unprecedented levels of individual and aggregate capacities in the order of Gbps wireless data rates [4]. This is in contrast to current Wi-Fi technologies (maximum efficiencies of 0.5 bits/s/Hz for 802.11b and 2.7 bits/s/Hz for 802.11a/g). Radio spectrum is also getting increasingly deregulated [6], which can potentially lead to an abundance of bandwidth when spectrum is used (and reused) more efficiently and cooperatively. The capacity increase and efficiency are a direct result of being able to switch between different idle channels only for the period of usage.

Multihop (or mesh) and long-range wireless technologies can potentially enable pervasive broadband access, much like electricity [8]. Multihop wireless simplifies private network deployment and management for residential users. Such networks are already very common in the residential neighborhoods of many cities in the United States, as well as in other major cities worldwide.

Wireless Solutions

In multihop wireless access, a fixed wireless access point is typically mounted on the rooftop of the subscriber's home. Each access point creates a small wireless coverage area called a "hop" and acts much like a router on the Internet, automatically discovering neighboring access points and relaying packets across several wireless hops. Note that multihop wireless is different from ad hoc wireless in that the fixed access points used in multihop wireless are not battery-powered, provide a much more reliable connection between hops, and can use a less complex but more efficient packet routing protocol. With multihop wireless, subscribers can create local community networks on demand, allowing residences to communicate directly with neighbors and enabling broadband applications between homes (e.g., neighborhood watchdog applications, medical and emergency response tasks, etc). Thus, the ownership of the access network becomes distributed across the residences. Such decentralized topologies not only provide a degree of autonomy (just as network domains or autonomous systems do in the Internet today) but can also lower subscriber costs substantially through shared services and resources. In addition, PTP traffic traveling within a neighborhood can be localized and distributed efficiently. While there is a need for innovative approaches to digital rights management (which will ultimately help to increase the availability of high-fidelity multimedia content), the ability to share resources is a key driver in reducing network service costs. We have seen the economies of scale provided by enterprise LANs in the 1980s, the public Internet in the 1990s, and more recently, PTP file sharing applications such as Kazaa (a

dominant on-line file-swapping platform) that has forced big recording labels to agree to offer music downloads for usually about $1 a song.

Other key advantages of operating distributed multihop wireless systems are lower transmit power (since there is no need to transmit information all the way back to the ultimate destination, which also results in a corresponding reduction in interference) and the ability to reuse limited radio spectrum efficiently. These advantages help increase the capacity of the wireless access network and improve range performance. For example, the network coverage of Wi-Fi can be extended in this manner and may ultimately play a part in reducing the costs for deploying residential broadband access networks.

Although multihop wireless provides unique benefits, it also presents tough challenges in guaranteeing security and QoS associated with always-on connections [1]. Overlay solutions involving commercial long-range Wi-Fi products can be deployed for improved service availability. These products provide an omnidirectional five-mile radius coverage using sensitive antenna arrays mounted on fixed locations and use only tens of milliwatts of transmit power (*Figure 3*). Base stations that service several miles can be located on a tower, a rooftop of a tall building, or other elevated structure. If directional high-gain antennas are used, then the operational range can extend up to 30 miles. Another popular long-range wireless solution is based on the IEEE 802.16 standard (www.ieee802.org/16), commercially known as WiMAX. More details are provided in [2].

Peer-to-Peer Content Download

While there has been much debate about the killer application for broadband access, possibly the most important technology that can allow any future killer application to happen is the downloading of content. The number of applications that can leverage on an improved downloading mechanism is virtually limitless. For example, cost-efficient next-generation phone systems may take the shape of decentralized, PTP file-sharing networks running over the Internet and eliminate phone-company middlemen [3]. The free Skype software program (www.skype.com) represents a big step in that direction and is challenging traditional VoIP. Skype allows users of the same application to make unlimited, high-quality, encrypted voice calls over the Internet and works with all firewalls, network address translators (NATs), and routers. With Skype, no VoIP media gateways are required since the PTP application effectively allows a voice packet to be downloaded quickly by the recipient from multiple sources. The success of PTP applications such as Skype is also boosted by a recent FCC ruling in February 2004 that voice communications flowing entirely over the Internet (i.e., voice services that do not interconnect the telephone system) are not subject to traditional government regulations and taxes applied to public switched telephone networks (PSTNs).

Peer-to-Peer File Sharing and Real-Time Streaming

Peer-to-peer file sharing accounts for more than 80 percent of Internet backbone traffic and overcomes the current lack of IP multicasting support by major Internet service providers (ISPs). Like wireless for the last mile, it is a disruptive but promising technology. In PTP communications,

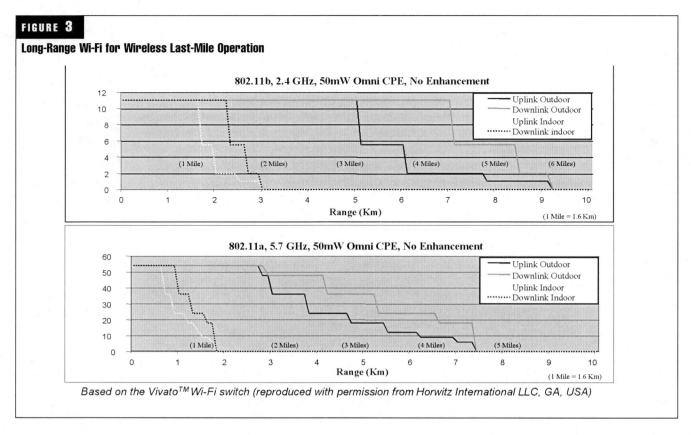

FIGURE 3

Long-Range Wi-Fi for Wireless Last-Mile Operation

Based on the Vivato™ Wi-Fi switch (reproduced with permission from Horwitz International LLC, GA, USA)

there are essentially two kinds of network nodes. Peers are computers that download and upload data. Trackers are servers assisting peers to find other peers. To demonstrate the potential of PTP performance gains, we use BitTorrent (a leading PTP distribution protocol) to run PTP file sharing and PTP radio to run real-time streaming over a data over cable service interface specifications (DOCSIS) 2.0 cable network (*Figures 4* and *5*). DOCSIS 2.0 is the latest cable standard (issued by CableLabs) to be commercialized. The standard provides symmetrical upstream and downstream data rates, which suits PTP applications better than asymmetrical data links, since each peer can act as a client and server. BitTorrent employs two kinds of files, namely, a data file and a small file that provides tracker location and data file description (e.g., data file length, piece identification, and hashing information). The basic mechanisms are as follows:

- *Publish*: Generate BitTorrent file and run tracker server
- *Join*: Contact tracker server, obtain list of peers
- *Piece*: Break data file into smaller pieces of fixed size, each downloaded piece is reported by all participating peers
- *Piece selection*: Rarest first, if not available, then random first
- *Fetch*: Download file pieces from peers, upload file pieces you have to peers

In *Figure 4*, background traffic on the backbone network and on the upstream link is generated using the Smartbits traffic generator. The average throughput per client is measured. In both experimental setups, the measured results show that PTP file distribution achieves higher throughput than the file transfer protocol (FTP) when the background traffic or

the number of peers increases. This is because information can be downloaded from multiple peers as opposed to one server in FTP. The throughput improvement can potentially be enhanced further using multihop wireless as detailed in Integrated Broadband Access Architecture. In *Figure 5*, good quality PTP streaming was achieved using 400 Kbps nullsoft video that is captured real-time using a video camcorder. Voice information was also captured real-time using a microphone and synchronized with the video transmission. The traffic map indicates the different peers exchanging information, while the monitor shows the data pieces kept by a specific peer.

Integrated Broadband Access Architecture

Currently, the prevalent model of an access network is that of a star topology with a hub at a central office location, and individual homes connected to the central office with an access line. Since all traffic needs to be directed to the central office, such a model creates a local bottleneck, especially when running PTP applications. An integrated multihop wireless and WDM PON architecture (*Figure 6*) can solve this problem with both networks acting as network overlays for improved reliability and performance. Besides providing a good alternative, such an architecture also represents an excellent evolutionary path for current access technologies (e.g., cable, DSL). The multihop wireless network achieves direct connectivity between local subscribers, can be expanded or reduced on demand, and allows PTP traffic traveling within a neighborhood to be localized (bypassing the central office) while ensuring all other incoming and outgoing PTP traffic from the central office to be distributed among these subscribers. Cooperative, distributed protocols

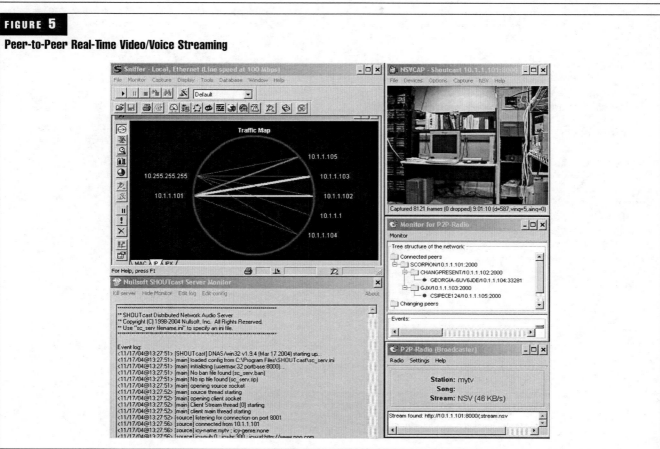

FIGURE 4

Peer-to-Peer File Sharing Performance

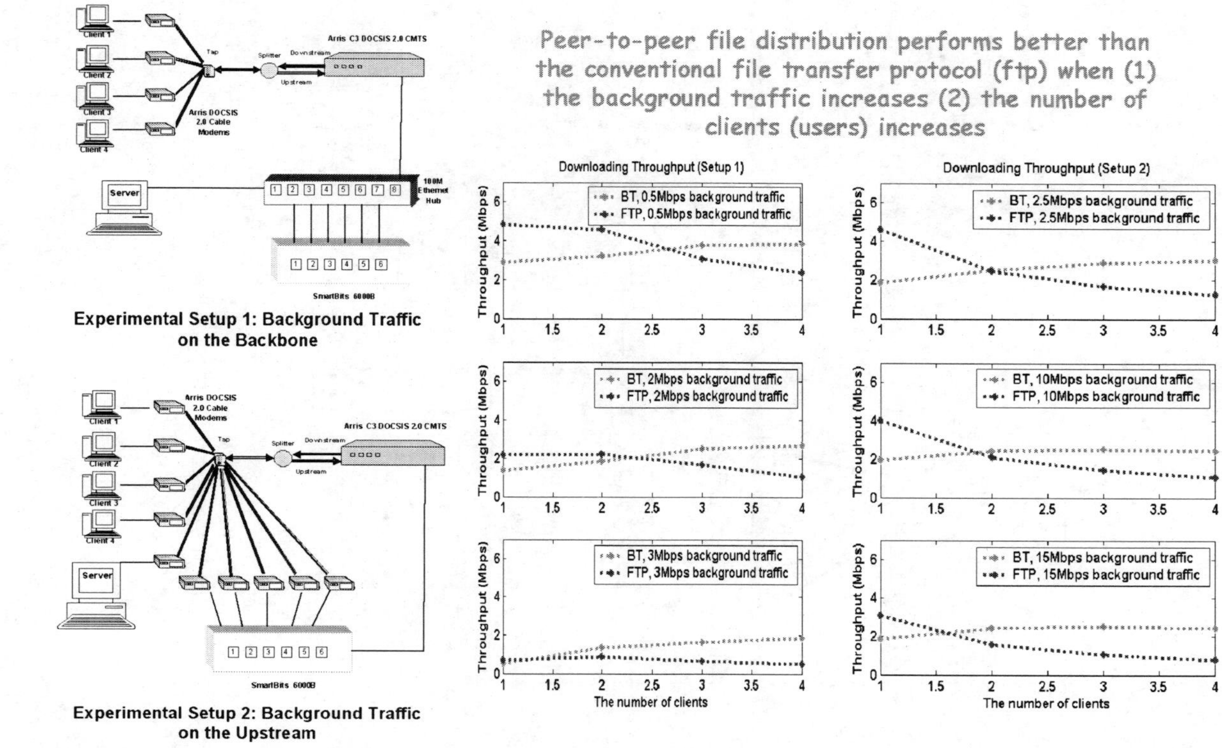

Peer-to-peer file distribution performs better than the conventional file transfer protocol (ftp) when (1) the background traffic increases (2) the number of clients (users) increases

Experimental Setup 1: Background Traffic on the Backbone

Experimental Setup 2: Background Traffic on the Upstream

FIGURE 5

Peer-to-Peer Real-Time Video/Voice Streaming

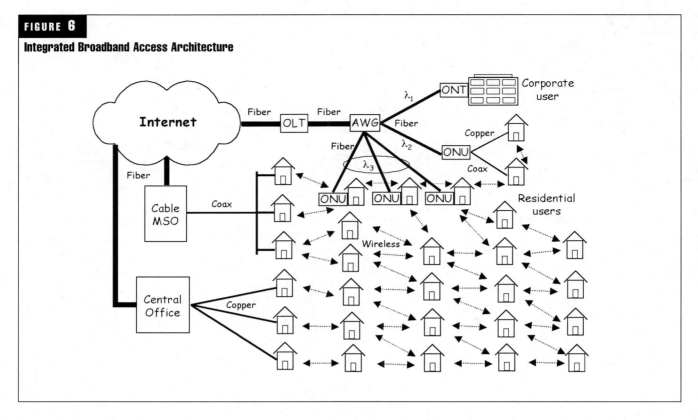

FIGURE 6

Integrated Broadband Access Architecture

are needed in such an integrated architecture. To provide more bandwidth needed for PTP and real-time applications, the wireless network may require intelligent protocols that facilitate radio spectrum reuse as well as opportunistic spectrum harvesting for improved efficiency, range, and network management, and increased bandwidth. The WDM PON reservation protocol with preallocated bandwidth discussed in A WDM Broadband Optical Access Protocol plays an important role in optimizing the delay performance of PTP file sharing and real-time streaming applications.

Conclusions

The emergence of distributed, autonomous topologies and PTP content download can help ensure that broadband access technologies are available to all for broad use and will ultimately lead to innovative applications. By combining the strengths of wireless and optical access technologies with symmetrical upstream and downstream data rates, pervasive broadband access and seamless integration with PTP applications can be achieved.

References

NSF Workshop Report, "Residential Broadband Revisited: Research Challenges in Residential Networks, Broadband Access, and Applications," January 20, 2004.

B. Bing, All in a Broadband Wireless Access Network: A Comprehensive Guide to the Next Wireless Revolution, December 2005, ISBN: 0976675218.

"Net Telephony as File-Trading," Business Week Online, January 6, 2004.

N. Jayant, "Scanning the Issue: Special Issue on Gigabit Wireless," Proceedings of the IEEE, Vol. 92, No. 2, February 2004, pp. 1–3.

Jey K. Jeyapalan, "Municipal Optical Fiber through Existing Sewers, Storm Drains, Waterlines, and Gas Pipes May Complete the Last Mile," www.agc.org/content/public/PDF/Municipal_Utilities/Municipal_fiber.pdf.

G. Staple and K. Werbach, "The End of Spectrum Scarcity," IEEE Spectrum, March 2004, Vol. 41, No. 3, pp. 49–52.

C. Xiao, B. Bing, G. K. Chang, "An Efficient Reservation MAC Protocol with Preallocation for High-Speed WDM Passive Optical Networks," 24th IEEE INFOCOM, March 2005.

R. Yassini, Planet Broadband, Cisco Press, 2004.

High-Power Lithium Ion

A New Area in Portable Power

Isidor Buchmann

President
Cadex Electronics, Inc.

Until recently, applications for high-current-rate capabilities were reserved for nickel-cadmium and nickel-metal-hydride batteries. These applications include power tools and medical equipment. Much has changed in the fields of lithium-based batteries in the past few years, and high-current lithium-ion now claim capabilities similar to or higher than nickel-based systems. These new-technology batteries are expected to have a similar impact on high-power portable products as the introduction of lithium-ion had on the consumer electronics market in the 1990s.

In spite of its great popularity, the cobalt-based lithium-ion has some limitations. It is not very robust and cannot take high charge and discharge currents. Trying to force a rapid charge or loading the battery with excess discharge current would overheat the pack and its safety would be jeopardized. The safety circuit of the cobalt-based battery is typically limited to a charge and discharge rate of about 1C. This means that a 2400 milliampere-hours (mAh) 18650 cell can only be charged and discharged with a maximum current of 2.4 A. Another drawback of the cobalt system is the increase of the internal resistance that occurs with cycling and aging. After two or three years of use, the pack often becomes unserviceable due to a high-voltage drop under load that is caused by elevated internal resistance. This condition cannot be reversed.

New cathode material opened the door for higher rate capability. Moving away from cobalt also helped in lowering manufacturing costs. In 1996, scientists succeeded in using lithium manganese oxide as a cathode material. This substance forms a three-dimensional spinel structure that provides improved ion flow on the electrode. High ion flow reflects in a lower internal resistance and hence higher loading capability. Unlike the cobalt-based lithium-ion, the resistance stays low with cycling and aging. The battery does age, however, and the overall service life is similar to that of the cobalt system. A further advantage of spinel is its inherent high stability. Spinel needs less in terms of safety circuit compared to the cobalt system.

Low internal cell resistance is the key to high rate capability. This characteristic benefits both in fast-charging and high-current discharging. A spinel-based lithium-ion in an 18650 package can, for example, be discharged at currents of 20 to 30 A with marginal heat buildup. One-second load pulses of twice the specified current are permissible. At continuous high-load requirements, a heat build-up will occur, and a cell temperature cannot exceed 80∞ C. Beside power tools and medical instruments, the spinel-based lithium-ion is a candidate for hybrid cars. Manufacturing cost will need to be lowered and the service life prolonged before this battery system can be used for automotive propulsion applications.

The spinel battery has some disadvantages, however. One of the largest drawbacks is the lower capacity compared to the cobalt-based system. Spinel provides roughly 1,200 mAh in an 18650 package, about half that of the cobalt equivalent. In spite of this, spinel provides an energy density that is about 50 percent higher than that of a nickel-based equivalent.

Types of Lithium-Ion Batteries

Lithium-ion has not yet reached full maturity, and the technology is continually improving. The anode in today's cells is made up of a graphite mixture, and the cathode is a combination of lithium and other choice metals. It should be noted that all materials in a battery have a theoretical energy density. With lithium-ion, the anode is well optimized and little improvements can be gained in terms of design changes. The cathode, however, shows promise for further enhancements. Battery research is therefore focusing on the cathode material. Another part that has potential is the electrolyte. The electrolyte serves as a reaction medium between the anode and the cathode.

The battery industry is making incremental capacity gains of 8 to 10 percent per year. This trend is expected to continue. This, however, is a far cry from Moore's Law, which specifies a doubling of transistors on a chip every 18 to 24 months. Translating this increase to a battery would mean a doubling of capacity every two years. Instead of two years, lithium-ion has doubled its energy capacity in 10 years.

Today's lithium-ion comes in many "flavors," and the differences in the composition are mostly related to the cathode material. *Table 1* below summarizes the most commonly used lithium-ion on the market today. For simplicity, we summarize the chemistries into four groupings, which are cobalt, manganese, nickel-cobalt manganese (NCM), and phosphate.

FIGURE 1

Most Lithium-Ion Batteries Used for Portable Applications Are Cobalt-Based. High Production Volume Enables the Battery to Be Manufactured at a Relatively Low Cost. They Consist of a Cobalt Oxide in the Positive Electrode (Cathode) and a Graphic Carbon in the Negative Electrode (Anode). One of the Main Advantages of the Cobalt-Based Battery Is Its High Energy Density. This Makes This Chemistry Attractive for Cell Phones, Laptop Computers, and Cameras.

The cobalt-based lithium-ion appeared first in 1991, introduced by Sony. This battery chemistry gained quick acceptance because of its high energy density. Possibly due to lower energy density, spinel-based lithium-ion had a slower start. When introduced in 1996, the world demanded longer run time above anything else. With the need for high current rate on many portable devices, spinel has now moved to the frontline and is in hot demand. The requirements are so great that manufacturers producing these batteries are unable to meet the demand. This is one of the reasons why so little advertising is done to promote this product. E-One

Moli Energy (Canada) is a leading manufacturer of the spinel lithium-ion in cylindrical form. They are specializing in the 18650 and 26700 cell formats. Other major players of spinel-based lithium-ion are Sanyo, Panasonic, and Sony.

Sony is focusing on the NCM version. The cathode incorporates cobalt, nickel, and manganese in the crystal structure that forms a multi-metal oxide material to which lithium is added. The manufacturer offers a range of products within this battery family, catering to users that either need high energy density or high load capability. It should be noted that these two attributes could not be combined in one and the same package; there is a compromise between the two. Note that the NCM charges to 4.10 V/cell, 100 mV lower than cobalt and spinel. Charging this battery chemistry to 4.20 V/cell would provide higher capacities, but the cycle life would be cut short. Instead of the customary 800 cycles achieved in a laboratory environment, the cycle count would be reduced to about 300.

The newest addition to the lithium-ion family is A123 Systems, which is adding nano-phosphate materials to the cathode. Although the manufacturer has not officially announced what metal is being used, it is widely believed to be iron. They claim to have the highest energy density of a commercially available lithium-ion battery. The cell can be continuously discharged to 100 percent depth-of-discharge at 35 C and endures discharge pulses as high as 100 C. The phosphate-based system has a nominal voltage of about 3.25 V/cell. The charge limit is 3.60 V. This is far lower than the customary 4.20V/cell of the cobalt-based lithium-ion. Because of these lower voltages, A123 Systems' batteries will need to be charged with a special charger. Due to the anticipated strong demand, this cell is expected to be in short supply. A123 Systems was founded in 2001, is privately held, and has major shareholders that include Motorola, QUALCOMM, and MIT.

TABLE 1

Most Common Types of Lithium-Ion Batteries

Chemistry	Nominal V	Charge V limit	Charge & discharge C rates	Energy density (Wh/kg	Applications	Note
Cobalt	3.6	4.20	1 limit	110–190	Cell phone, cameras, laptops	Since the 1990s, most commonly used for portable devices; has high energy density
Manganese (spinel)	3.7–3.8	4.20	10 cont. 40 pulse	110–120	Power tools, medical equipment	Low internal resistance; offers high current rate and fast charging, but lower energy density
NCM (nickel-cobalt manganese)	3.7	4.10*	~5 cont. 30 pulse	95–130	Power tools, medical equipment	Nickel, cobalt, manganese mix; provides compromise between high current rate and high capacity
Phosphate (A123 System)	3.2–3.3	3.60*	35 cont.	95–140	Power tools, medical equipment	New, high current rate, long cycle life; higher charge V increases capacity but shortens life cycle

* = Higher voltages provide more capacity but reduce cycle life

Confusion with Voltages

For the past 10 years or so, the nominal voltage of lithium-ion was known to be 3.60 V/cell. This was a rather handy figure because it made up for three nickel-based batteries (1.2 V/cell) connected in series. Using the higher cell voltages for lithium-ion reflects in better watt-hour readings on paper and poses a marketing advantage; however, the equipment manufacturer will continue assuming the cell to be 3.60 V.

The nominal voltage of a lithium-ion battery is calculated by taking a fully charged battery of about 4.20 V, fully discharging it to about 3.00 V at a rate of 0.5 C while measuring the average voltage. Because of the lower internal resistance, the average voltage of a spinel system will be higher than that of the cobalt-based equivalent. Pure spinel has the lowest internal resistance, and the nominal cell voltage is 3.80 V. The exception again is the phosphate-based lithium-ion. This system deviates the furthest from the conventional lithium-ion system.

Prolonged Battery Life through Moderation

Batteries live longer if treated in a gentle manner. High-charge voltages, excessive charge rate, and extreme load conditions will have a negative effect and shorten the battery life. This also applies to the high current rate lithium-ion batteries. The longevity is often a direct result of the environmental stresses applied. The following guidelines suggest how to prolong battery life.

The time at which the battery stays at 4.20 V/cell should be as short as possible. Prolonged high voltage promotes corrosion, especially at elevated temperatures. Spinel is less sensitive to high voltage.

3.92 V/cell is the best upper voltage threshold for cobalt-based lithium-ion. Charging batteries to this voltage level has been shown to double cycle life. Lithium-ion systems for defense applications make use of the lower-voltage threshold. The negative is reduced capacity.

The charge current of lithium-ion should be moderate (0.5 C for cobalt-based lithium-ion). The lower charge current reduces the time in which the cell resides at 4.20 V. It should be noted that a 0.5 C charge only adds marginally to the charge time over 1 C because the topping charge will be shorter. A high current charge tends to push the voltage up and forces it into the voltage limit prematurely.

A depth of discharge of 80 percent or less poses less strain on the battery than a full 100 percent discharge. It is better to charge lithium-ion more often than letting it down too deeply. One does not need to worry about memory, as was the case with nickel-cadmium.

Note: In respect to fast charging and topping charge, the charge behavior of lithium-ion is similar to lead acid. Here, the voltage threshold of 2.35 V/cell during regular charge needs to be lowered to 2.27 V/cell when the VRLA is on standby. Keeping the voltage at the high threshold would contribute to corrosion. A similar effect occurs with lithium-ion.

Not only does a lithium-ion battery live longer with a slower charge rate, high discharge rates also contribute to the extra wear and tear. *Figure 2* shows the cycle life as a function of charge and discharge rates. Observe the good laboratory performance if the battery is charged and discharged at 1 C. (A 0.5 C charge and discharge would further improve this rating.)

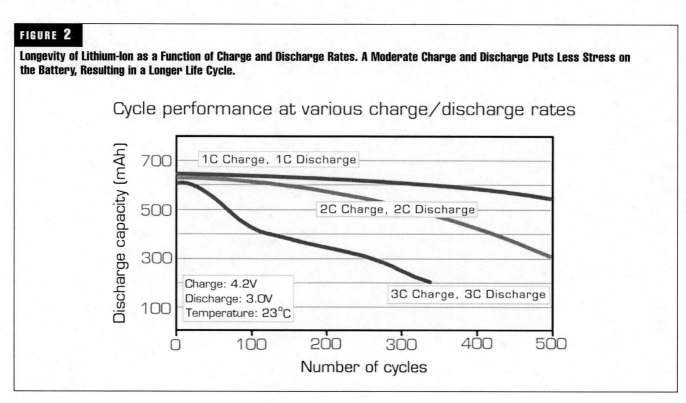

FIGURE 2

Longevity of Lithium-Ion as a Function of Charge and Discharge Rates. A Moderate Charge and Discharge Puts Less Stress on the Battery, Resulting in a Longer Life Cycle.

Battery experts agree that the longevity of lithium-ion is shortened by other factors than charge and discharge rates alone. Elevated temperature while the battery is fully charged is one such element that shortens service life. Even though incremental improvements can be achieved with careful use of the battery, our environment and the services required are not always conducive to achieve optimal battery life. In this respect, the batteries behave much like us humans—we cannot always live a life that is favorable of achieving a maximum life span.

Profiting from Convergence

Defining Growth Paths for Telecom Service Providers

Jeanette Carlsson

Global Communications Sector Leader
IBM Institute for Business Value

Chris Woodland

Senior Telecom Industry Consultant
IBM Business Consulting Services

Henry Stevens

Global Wireless Strategy and Solutions Leader
IBM Business Consulting Services

Introduction

Internal and external forces of convergence are taking hold of the telecommunications industry. Convergence of local and national voice services dramatically changed the telecom landscape, as competition flourished. Now, voice and data services are rapidly converging. Soon, customers will be able to access any content or application seamlessly from a multitude of networks, using any device of their choosing.

Industry players are gearing up to harness the potential of converging technology, networks, devices and content to develop multimedia services and solutions of ever-increasing sophistication on a single Internet Protocol (IP). Evolving customer demands for content from an increasing variety of sources will require telecom providers to engage in a complex web of collaboration with the media and entertainment, IT and consumer electronics industries.

Convergence offers telecom companies a world of opportunities. However, before they can profit, telecom providers must make tough strategic and technical choices. Success will depend not only on making the right decisions about where to play in a converged environment, but on transforming business models and capabilities to make the most of these new opportunities. Most importantly, service providers must align their strategic choices and capabilities to the specific needs of the customer segments they intend to serve. The days of being "all things to all people" are in the past.

On the verge of opportunity

Several market factors are reshaping the telecom industry over the next decade:

- Intense competition is driving voice prices down: between 2003 and 2006, revenue growth for mobile providers could decline by more than half, 1 while current growth for many fixed line businesses is flat or even negative.

- Consumer spending on broadband is expected to rise 37 percent, to over US $100 billion by 2007.[2]

- Music and gaming content will explode: consumer expenditures in these areas is anticipated to increase 185 percent and 78 percent, respectively, over the same time frame.[3]

- Device and IP network proliferation continues: the last four years have seen a 100 percent increase in smart phone sales and international Voice over IP (VoIP) minutes.[4]

- Rapid emergence of new points of high-speed Internet access: global wireless fidelity (Wi-Fi) hotspots totaled 49,700 in 2003 and are expected to total 190,000 by 2007.[5]

Together, these factors are driving the convergence of telecommunications with other industries, and creating unprecedented change and growth potential for telecom providers as traditional "product markets" decline and new service opportunities arise.

This growth in supply and demand for new technologies and services facilitated by IP technology will blur the traditional boundaries of service, device and network—giving rise to a new "converged ecosystem" where telecom companies must partner to create value. As the single-product, voice-only world of telecom evolves into a multifaceted services industry, simple connectivity will be increasingly commoditized. Value will be created by providing services and solutions to consumer, enterprise and public sector customers.

Where telecom network operators have historically controlled all elements of the value chain, the introduction of new players in more complex value systems is making it increasingly difficult to build lasting relationships with customers. In this environment, telecom providers must make clear decisions about what kind of businesses they want to be. There will be both winners and losers in all major telecom provider categories. Success will depend on each player's ability to combine its own differentiating strengths (network assets, customer management, service creation, etc.) with the capabilities of partners to create seamless communications services that meet the needs of targeted customer segments.

To understand just how much impact convergence will have on the telecom industry, the IBM® Institute for Business Value sponsored an economist intelligence unit (EIU) survey of top industry executives and assessed the current market factors affecting the industry. The resulting study identifies the top-of-mind concerns for industry executives, and outlines the key challenges for telecom companies as convergence begins to take hold.

Competing in the Era of Convergence

Before convergence: Cracks in connectivity

Charles, a sales director for a European media firm, begins a typical workday at 7:00 a.m. at his home in Amsterdam, booting up his laptop. This morning, he must download the latest sales presentation so that he can update it while traveling to London for the monthly Board meeting this afternoon. After five minutes of booting up and dialing in to the network, a familiar message appears: the server is down. If he waits to retry, he will be late for his flight. Instead, he leaves a voicemail with a colleague, asking for an update to be sent to his BlackBerry hoping that he can pick up the new presentation on the way.

By 9:00 a.m., Charles is on the train to the airport. Although the trip gives him time to peruse the morning paper, the presentation—now just hours away—weighs heavily on his mind. He wishes the train offered broadband coverage so he could make better use of the time. He thinks he might have a chance to download the presentation in the airport lounge where public wireless LAN is available. But when he arrives at the airport, the security line is excruciatingly long, and again a chance to download the presentation eludes him.

On the London Heathrow Express transit, Charles receives a BlackBerry e-mail: the latest slides are now in his work e-mail inbox. Better late than never. He jumps into a taxi. As he speeds toward his meeting, he watches the new in-cab TV service's financial news update. The dollar has fallen further on the euro which is bad news for next quarter's sales prospects.

Charles arrives at the meeting venue and, with just thirty minutes to go, he races to find a spare desk and Ethernet cable to download the new presentation. The file is a whopping 20 MB, and a few slides need to be amended. In the end, the meeting is not particularly good: he is 10 minutes late and unprepared to answer several tricky questions on the new numbers in the updated presentation.

At 4:00 p.m., it's straight back to the airport, even without time to check and reply to e-mail or get updates on the several other projects he has been disconnected from all day.

Predictably, the plane sits on the runway for an hour before it departs. It is 8:00 p.m. before Charles returns home again, tired and irritable. He can't help but think there must be a better way to use his time, manage his work day and his life, for that matter.

Eighty percent of the telecommunications executives surveyed agreed that it is essential to embrace convergence within the next three years as a source of long-term revenue growth.[6] Moreover, there was a clear correlation between the perceived importance and the likely timing of different types of convergence.

Those surveyed view voice and data as the most important type of convergence (see *Figure 1*),[7] as evidenced by the strong growth in VoIP in all geographic markets today. Service providers are able to offer VoIP at highly competitive prices, without owning network assets and avoiding all the traditional constraints of distance, location and unit pricing models.

Most incumbent fixed network operators have overcome their initial fears of revenue cannibalization. As revenues from fixed line voice minutes decrease, it is "eat or be eaten." Many are launching their own VoIP services, bundled with digital subscriber line (DSL) broadband subscriptions as a value-added service.

Telecom executives perceive convergence between fixed line and wireless access technologies to be the next most important type of convergence to impact their business.[8]

As customers grow accustomed to the increased bandwidths of their fixed home, office and vehicular networks—and the ever-presence of their mobile (cellular) networks—combining the two will become increasingly attractive. Fixed line service providers are targeting this opportunity by bundling WiFi access with DSL subscriptions, while mobile service providers are offering integrated 2.5G/3G/WiFi PC data cards to accomplish the same.

The key to this area of convergence is wireless (but not mobile) technologies that can "bridge" between fixed line and mobile networks: hence the success of WiFi (802.11x) and Bluetooth, and the interest in the evolution of wireless technologies such as WiMax (802.16x), ultrawideband (UWB), near field communications (NFC) and Zigbee. Eventually, this type of convergence is expected to involve accessing fixed, wireless, and mobile networks seamlessly with a single device to create a truly converged service, as envisioned by BT's Project Bluephone and others.

The forces of convergence discussed above are primarily internal to the telecommunications industry, and it is not surprising that our executive survey group today regards them as more important, since telecom providers are still able to retain a relatively high degree of control over their development. However, the real challenges for the telecom industry are external, emanating from the impact of convergence with the IT services, media and entertainment, and consumer electronics industries.

Some players in these industries—with well-established brands, customer bases, distribution channels and competencies—see telecommunications as the next logical exten-

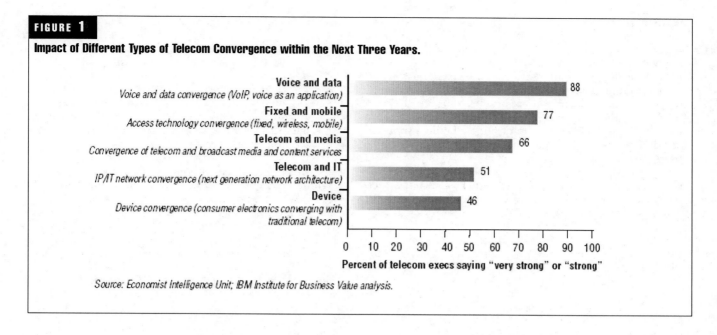

FIGURE 1

Impact of Different Types of Telecom Convergence within the Next Three Years.

Source: Economist Intelligence Unit; IBM Institute for Business Value analysis.

sion of their businesses. For telecom providers, this poses unique challenges in terms of customer relationships, service development, next generation network (NGN) development, and the systems and processes required to deliver working solutions to customers.

Challenge: Customer focus
Forty-six percent of the executives that responded to the EIU study stated that their medium-term growth strategies centered on core markets and customers.[9] But, how well do today's service providers know what tomorrow's customers will require? In a converging environment, the ways in which consumers live, work and communicate at home, in the business world and even in their cars, are changing rapidly.

Developing greater customer intimacy can help telecom service providers differentiate the customer experience, as well as stimulate usage and loyalty. Implementing needs-based segmentation, investing in skills, capabilities and systems, and enhancing market research can help operators make strides in this area. Improving the customer experience requires attention across all stages of the customer journey, including awareness, acquisition, growth and retention (see *Figure 2*).

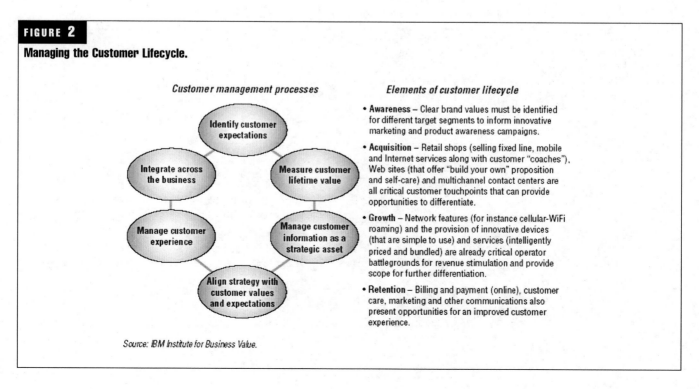

FIGURE 2

Managing the Customer Lifecycle.

Customer management processes

Identify customer expectations

Integrate across the business

Measure customer lifetime value

Manage customer experience

Manage customer information as a strategic asset

Align strategy with customer values and expectations

Elements of customer lifecycle

• **Awareness** – Clear brand values must be identified for different target segments to inform innovative marketing and product awareness campaigns.

• **Acquisition** – Retail shops (selling fixed line, mobile and Internet services along with customer "coaches"), Web sites (that offer "build your own" proposition and self-care) and multichannel contact centers are all critical customer touchpoints that can provide opportunities to differentiate.

• **Growth** – Network features (for instance cellular-WiFi roaming) and the provision of innovative devices (that are simple to use) and services (intelligently priced and bundled) are already critical operator battlegrounds for revenue stimulation and provide scope for further differentiation.

• **Retention** – Billing and payment (online), customer care, marketing and other communications also present opportunities for an improved customer experience.

Source: IBM Institute for Business Value.

Customer segments are quickly evolving as customer needs become more distinct. Service provider offerings must match the connectivity needs of their customers' lifestyles. For instance, consumers of different age groups interact and communicate for different reasons (see *Figure 3*). Understanding not only how these consumer groups value, access, and use content and applications, but how to build new propositions that target these users, will be critical to competitiveness. Consumer propositions should also reflect key differences in the needs of fixed-broadband based home users (leveraging DSL/cable television [CATV] and WiFi home networks) and mobile personal users that require seamless roaming between cellular and fixed wireless networks whether they are at home, at the office, in the car or simply walking down the street.

Small and medium-sized enterprises (SMEs) represent a lucrative opportunity for telecom companies. However, the disparity of SME size, revenue potential and technical expertise make it more difficult for large telecom providers to sell directly to this segment than to large enterprises. SMEs require reliability, flexibility of service and responsiveness to their business needs. Value for money is more important than simple low costs. Above all, SME customers want to be treated in a manner that reflects the reliance they have on telecom services for their own businesses to succeed.

To successfully develop the SME market opportunity, telecom providers need to keep offerings simple and streamlined, with end-to-end customer service and experiences tailored to SME needs. Stripped-down large enterprise solutions and repackaged offerings will no longer suffice. SMEs are very close to their own customers, and want the same from their suppliers: telecom marketing, sales, and service must all connect with SMEs in ways that reflect how SMEs do business.

In contrast, many big businesses look to telecommunications as a necessary business tool, but also as an opportunity for reducing cost. In this context, the telecom industry must shift the focus from cost reduction to productivity improvement in order to succeed.

Convergence between telecom and IT services offers a very significant opportunity to develop applications targeted to the large enterprise that use converged products and services to increase efficiency and generate savings. Indeed, value from converged services must be created from the business impact of the service itself, not just the simple act of transferring voice or data from one device to another.

Finally, service providers are likely to use converged services in interactions with their customers. Significant increases in self-service and speech-enabled technologies are forecast in contact center environments, combining voice and data to give customers greater choice in how they request and receive care, billing and other information from service providers. Trials to date have delivered substantial cost savings and significant increases in customer satisfaction.

Challenge: New service creation
Convergence enables growth through new services that benefit from converged technologies, increased customer

FIGURE 3

The Needs of Evolving Consumer Segments.

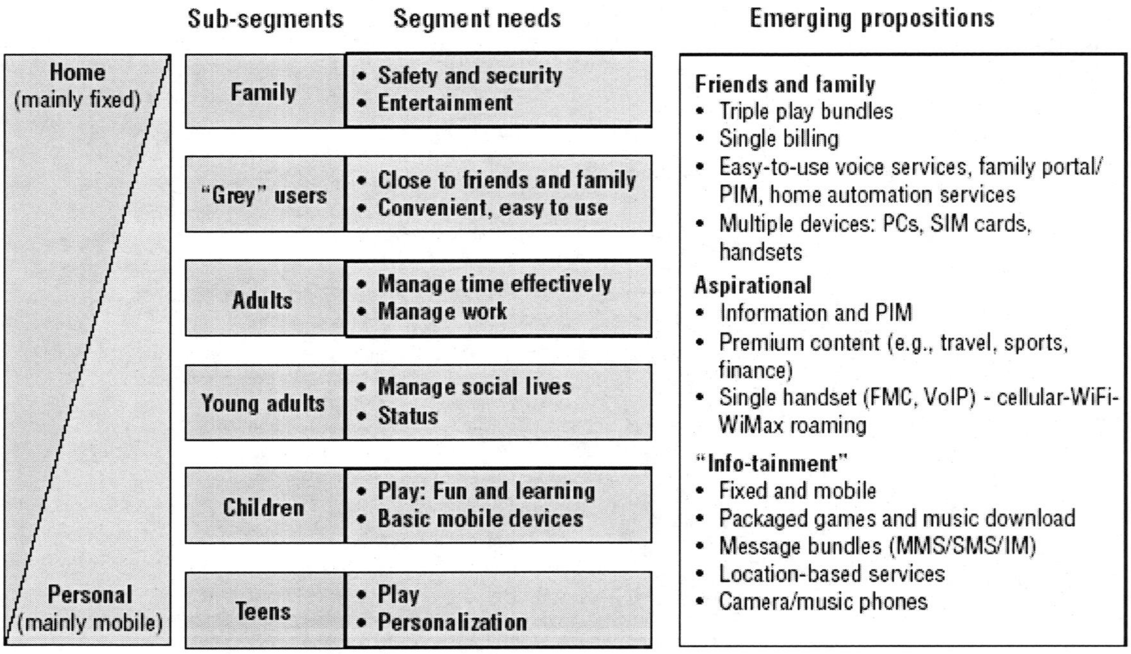

Source: IBM Institute for Business Value.

intimacy and third-party content. Telecom executives in the EIU survey overwhelmingly see new products and services and expansion in wireless access technology as key sources of medium-term revenue growth.[10] However, bringing complex converged services to market quickly depends on dramatically enhancing new product development processes. New services need to be innovative, customer-centric, and priced to drive adoption.

Telecom providers must look for ways to utilize core capabilities to enter new markets and service areas. Using existing customer footholds, they can expand product and service propositions to better target particular customer segments.

Challenge: Devices

Ironically, the telecom executives surveyed do not currently view device convergence as particularly important to creating value, in comparison to other industry developments.[11] This may be because many of the respondents have a traditional view that maps service rollout to device capability, and also because devices are often an area of the telecom value chain over which service providers do not have the level of control they would like. Yet the success of new devices such as the Apple iPod is undeniable evidence that device developments can not only make the difference, but can actually be the primary driver of new service penetration – and can change the game quickly.

Next generation mobile phones are expected to bridge the gaps between networks and devices, enabling users to access their digital content anywhere, using the device of their choosing. Device functionality and affordability are improving and converged devices—terminals that combine various wireless technologies in one product—are now emerging. As more multimedia and enhanced-data capabilities—such as the ability to stream media—become standard additions to mobile devices, users will increasingly be able to communicate in various ways. Ovum forecasts an increase in penetration of "feature phones" (voice-centric devices with multimedia capability that run on open platforms such as Java®) and "smartphones" (feature phones that run a full operating system such as Symbian OS or Microsoft® OS) between 2003 and 2008 (see *Figure 4*).[13]

Putting together the jigsaw puzzle of emerging devices, software, and distribution systems will be a lucrative business. Only a tight ecosystem of content providers, hardware, software, and telecom partners will transform the current complexity into a simple user experience. NTT DoCoMo has led the way for mobile operators in seizing control over device specifications while working with manufacturing partners, closely followed by Vodafone, leveraging its global scale in this area. We anticipate that this trend will continue, as the device becomes increasingly important in service delivery and adoption.

The first multimode cellular voice over WiFi handsets are starting to appear. According to The Shosteck Group, 15,000 to 20,000 of these converged handsets could ship this year.[13] While early phones will likely resemble mid-tier voice phones, multimodal functionality is expected in smartphones, which are forecasted to experience strong sales growth in the coming years. Converged devices are

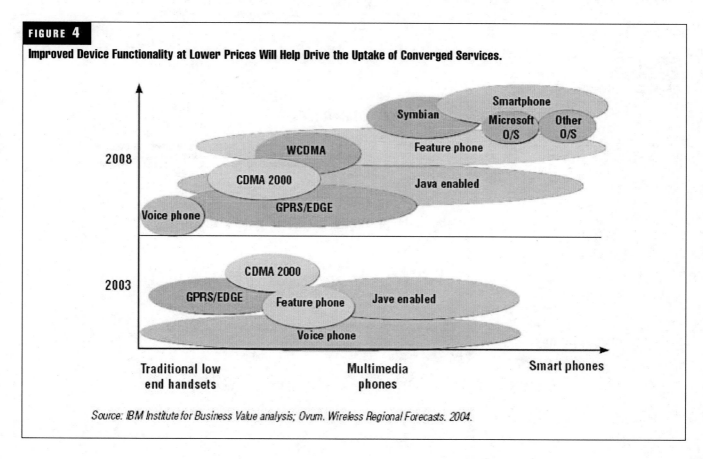

FIGURE 4

Improved Device Functionality at Lower Prices Will Help Drive the Uptake of Converged Services.

Source: IBM Institute for Business Value analysis; Ovum. Wireless Regional Forecasts. 2004.

expected to prosper as short-term issues, such as WiMax spectrum licensing, are resolved.

The battle to dominate the digital home depends on the proliferation of wireless networks that transmit entertainment and media within the living room, and from room to room. New devices and software to store content are expected to be plentiful, from game consoles to PCs, TVs, and other hybrid home gateways.

To date, only Microsoft has a comprehensive mix of software (including its Media Center PC with a special version of the Windows® operating system for handling digital content), hardware (the Xbox® games console), and consumer Internet services (MSN online music store).

Device convergence within vehicles is underway as well. An increasing proportion of high-end cars now feature wireless solutions that allow drivers to talk hands-free using a Bluetooth connection. In the future, cars may "talk" to one another using radio technologies, including WiFi. For instance, BMW has announced its "ConnectedDrive" solution, which could help alleviate long drive times through the use of WiFi. If caught in a traffic jam, the car can relay information to similarly equipped vehicles in the area, which then plot alternative routes to avoid the gridlock.[14]

Challenge: NGNs, operating support systems and business support systems (OSS/BSS)

As access and network technologies multiply, it becomes increasingly important for telecom companies to invest in networks that are access-technology agnostic. In a converged environment, success comes from an IP core that can interact with any sort of access technology to provide cus-

tomers with "anytime, anywhere" access to content, while allowing service providers to view and manage all customer information. An IP multimedia subsystem (IMS) enables operators to increase earnings before interest, taxes, depreciation, and amortization (EBITDA) performance, by driving revenue from converged voice and data services across multiple access technologies and by reducing cost (see *Figure 5*).

Internally, telecommunications providers must move away from vertical product silos to a horizontal focus that cuts across silos to include network, applications and customer management. Addressing the demand for simple, converged services requires enterprises to align around customer needs and avoid disjointed operations. Disjointed operations and redundant capabilities can result in inefficient cost structures and inconsistent customer experiences. NGNs provide a framework that enables service providers to migrate to an enhanced service delivery capability (including OSS/BSS billing support services, seamless roaming across cellular and fixed wireless access [FWA] networks and device management) and remove duplicate activities.

Forty-one percent of executives surveyed agree that providing reliable and affordable services will be a key characteristic of their growth strategy over the next few years.[14] However, most telecom providers are currently squeezed between pricing their offerings competitively and the high cost of non-integrated business processes. These pressures are compounded by multiplying product and service categories and increased customer volume. To top it all off, billing inaccuracies breed customer dissatisfaction and can result in lost revenues. Typically, telecom providers have attempted to solve these problems by consolidating applica-

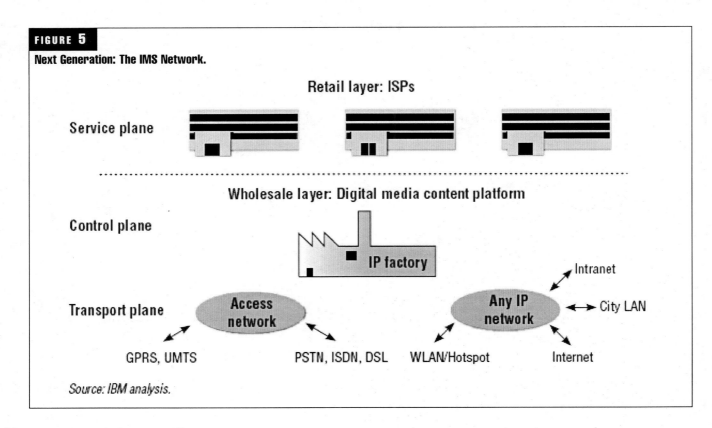

FIGURE 5

Next Generation: The IMS Network.

Retail layer: ISPs

Service plane

Wholesale layer: Digital media content platform

Control plane

IP factory

Transport plane

Access network

Any IP network

Intranet

City LAN

GPRS, UMTS   PSTN, ISDN, DSL   WLAN/Hotspot   Internet

Source: IBM analysis.

tions within individual functions or working to address IT costs rather than end-to-end business process costs. To date, initiatives to improve the customer experience have had limited success: high project failure rates, cost overruns, customer dissatisfaction and high churn rates plague many telecom providers.

Most telecom providers are still vertically aligned when it comes to billing as they have different billing systems for different services. Cutting through the resulting complexity of the different system types and layers is time consuming, delays time-to-market, and makes service bundling more difficult. Meanwhile, increasing complexity accelerates cost as companies struggle to keep pace with the market and offer new products and services. To become truly converged and reduce costs, telecom providers need to build one integrated, unified and automated billing system that supports all billing processes regardless of the product or service.

Making a play: Building converged business models
As the various forms of convergence take hold, established industry players face increasing competition, from each other as well as from insurgent business models such as virtual operators (e.g., mobile virtual network operators, VoIP service providers) and access utility players (e.g., FWA providers). Today's major players fall into four categories, fixed network operators with information, communications and technology (ICT) competencies, home broadband providers, integrated operators and mobile network operators. Each category must deal with their own challenges in building viable business models for the converged environment.

Convergence winners will be those companies within each category that leverage external reservoirs of rich content and applications in delivering well-packaged, focused propositions to customer segments, while embracing NGNs to streamline internal operations. Successful execution of convergence initiatives can drive significant EBITDA margin improvements when compared with business-as-usual (see *Figure 6*).

Fixed network operators and Information, Communications and Technology (ICT) providers
Companies such as BT, Cable & Wireless, MCI and AT&T have targeted the ICT market to create value beyond simple connectivity for enterprise customers, in addition to driving traffic over their high-capacity networks. We expect that many of these players will seek to provide full-service managed solutions, including mobility services, through wholesale capacity from other network owners.

To focus on providing "total solutions" to customers, these companies should:

- Work to develop NGNs that dynamically support new value-added services that improve the customer experience
- Form partnerships with companies that have strong system integration skills and assets

FIGURE 6

Upside of Convergence Case versus Baseline Case, 2008

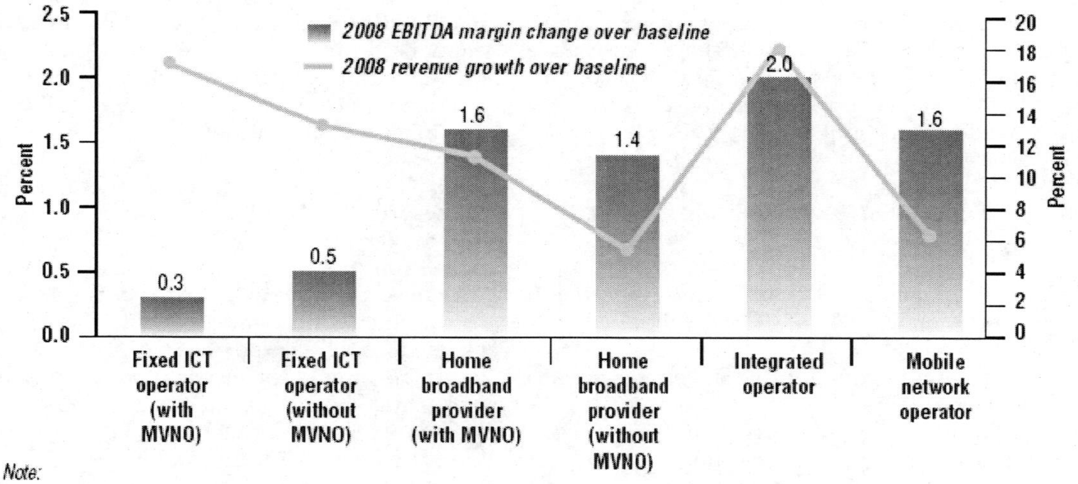

Note:
- Convergence initiatives vary by business model and include new revenue streams as well as the operating cost associated with generating new revenue (i.e., the cost of goods sold and other costs required for capability development).
- The fixed ICT convergence case includes core fixed line services, a share of ICT revenue and a share of mobile revenue (in the MVNO case).
- The home broadband provider case excludes the wholesale business, but retains core consumer fixed line services, including video-on-demand, VoIP, gaming, other media, fixed wireless access as well as revenue from a MVNO business (in the MVNO case).
- The integrated operator case includes core fixed and wireless businesses, including video-on-demand, VoIP, gaming, other media and fixed wireless access.
- The mobile network operator includes a fixed wireless access business in addition to its core cellular business (in which data usage is stimulated).

Source: IBM Institute for Business Value analysis.

- Develop more and varied retail distribution channels to target different industry verticals
- Build a wide geographic presence, potentially with partners, to provide global service to large corporations and multinational corporations (MNCs).

Home broadband providers

Cable companies and DSL providers dominate this space, delivering a triple play of voice, Internet and entertainment (possibly enriched by mobile services). These companies pose a significant threat to incumbent fixed network operators (DSL providers). Both cable and DSL providers face competition from new content providers such as AOL, Microsoft, and Yahoo!, with the result that most have either bought or developed capability in this area (e.g., Time Warner-AOL, France Telecom-Wanadoo) or allied with a major player (e.g., BT, SBC alliances with Yahoo!).

To drive further success, home broadband providers should:

- Use their anchor points in the residential market to deploy home gateway solutions
- Build and nurture their content relationships to drive service bundles
- Transition to all-digital broadband networks from narrowband
- Improve the customer experience by providing a simple and easy-to-use content navigation interface and seamless transfer to digital content between different devices (e.g., music from iPod to high fidelity [Hi-Fi])
- Investigate the benefits of a mobility play and fixed-mobile converged services (e.g., BT Bluephone).

Mobile network operators

The mobile pure-play is currently characterized by consolidation of competitive position, continued growth in mobile voice usage and revenues, and a strong desire to succeed in the more challenging area of mobile data. These companies have strong positions in the consumer market, focused in particular on individuals (as opposed to households), but are struggling to address the growing opportunity of enterprise mobile data services.

As they aim to offer ubiquitous bandwidth, these companies should:

- Build a flexible, open network platform designed to integrate new technologies more easily as they become available
- Create new pricing models to drive 3G/data penetration and service adoption
- Differentiate with an end-to-end customer experience that drives customer loyalty and revenue.

Integrated operators

This term is applied to those companies owning both fixed and mobile networks, although most such companies have not yet fully integrated the diverse parts of their businesses. These companies will be afforded perhaps the greatest opportunities from convergence, but also face the biggest hurdles to unlocking value, namely operational, regulatory and cultural challenges. To succeed, these operators will need to fully integrate not only their customer-facing processes, but also their network and their back office as well. They should:

- Drive the integration of fixed and mobile businesses and processes to reduce capital and operating expense (e.g., OSS/BSS, NGN)
- Deliver integrated customer experiences and services
- Develop integrated propositions to target the combined needs of both the individual and the household as a single customer (quadruple play)
- Investigate the outsourcing of network operations and maintenance.

Converged capabilities: Partnering to create value

Forty-six percent of executives surveyed stated that new products and services will primarily be developed in-house.[15] However, IBM believes that in a converged marketplace, telecom companies will increasingly need to look outside the enterprise to create differentiating products and services. Once telecom providers decide which business model and capabilities will help them make the most of convergence, they need to choose the strategic partners that can best help them deliver cutting-edge products and services (see *Figure 7*). Although building partnerships has not traditionally been a key focus for telecom companies, in a converged future, many may find that partnering is necessary to create value. Successful partnerships are an essential ingredient in allowing telecom providers to help satisfy customer needs and deliver a seamless customer experience.

What is your convergence quotient?

Completely converged: Cool and connected

At 7:00 a.m., our virtual worker Charles checks his home IPTV to see if he has an update of the latest sales presentation. But there is only a message from work: "Sorry, not quite finished yet, will be done within 3 hrs." Some things never change. However, Charles is confident he will be able to download the presentation on his way to London for this month's Board meeting. The airline he is flying has just announced that they have signed his telecom service provider as a partner for its in-flight high-speed wireless Internet service. Now, Charles can download the presentation to his laptop while in flight.

With a little time on his hands, Charles indulges himself. On the way to the airport, the on-train public WLAN allows him to pick up a video clip of the European football highlights from the night before on his "3G-WLAN" handset. Ajax of Amsterdam won, which puts him in a great mood. During the flight, Charles downloads the presentation as planned from his corporate e-mail account, without having to fumble around for the right currency to pay for it. The cost is automatically added to his work communications account. He has plenty of time to digest the new sales figures on the way to Heathrow.

When he arrives, Charles receives a video message from the London secretary advising him of a new meeting time and office venue. He is not too sure of this new location, but no problem—he has also been sent a multimedia map that he "Bluetooths" to his taxi driver's navigational screen. Charles makes it to the meeting venue with five minutes to spare. While waiting for the other executives to arrive, a multimedia message (MMS) notifies him that a new music album and video from his favorite artist has been released in the UK (ahead of the Netherlands)—does he want to preview and buy? He replies yes, but uses the voice-controlled

FIGURE 7

Telecom Providers Must Determine Where They Will Play a Full or Partial Role, and Which Non-Core Activities Should Be Outsourced to Third Parties.

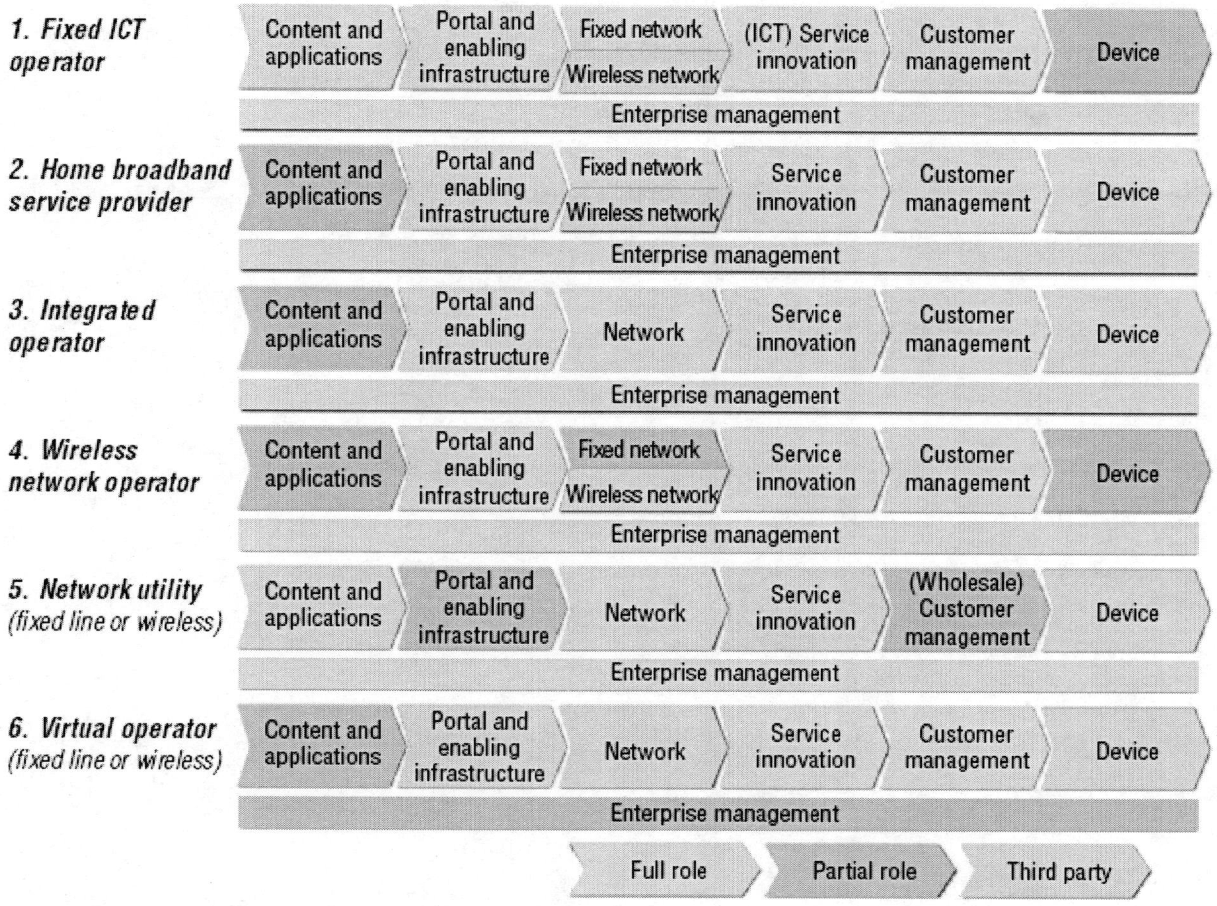

1. Fixed ICT operator — Content and applications | Portal and enabling infrastructure | Fixed network / Wireless network | (ICT) Service innovation | Customer management | Device
Enterprise management

2. Home broadband service provider — Content and applications | Portal and enabling infrastructure | Fixed network / Wireless network | Service innovation | Customer management | Device
Enterprise management

3. Integrated operator — Content and applications | Portal and enabling infrastructure | Network | Service innovation | Customer management | Device
Enterprise management

4. Wireless network operator — Content and applications | Portal and enabling infrastructure | Fixed network / Wireless network | Service innovation | Customer management | Device
Enterprise management

5. Network utility (fixed line or wireless) — Content and applications | Portal and enabling infrastructure | Network | Service innovation | (Wholesale) Customer management | Device
Enterprise management

6. Virtual operator (fixed line or wireless) — Content and applications | Portal and enabling infrastructure | Network | Service innovation | Customer management | Device
Enterprise management

Full role | Partial role | Third party

Source: IBM Institute for Business Value.

---

menu to redirect the track to his home DSL service so he can enjoy it later.

The meeting is successful, and Charles feels that he was well prepared and that the Board members are happy. It is 5:00 p.m. and time to get to the airport. In the taxi, Charles decides to check out that album. Now that the City of Westminster has deployed a metro WiMax network, there is no need to wait to access the music—and he'll have some well-earned entertainment for the flight home.

Is your company prepared to offer customers this kind of seamless connectivity to the content they value? Executives responding to the EIU study agreed that for them, profiting from convergence will first require the resolution of key demand, operational and supply challenges. Those surveyed saw competition (87 percent), regulation (79 percent), understanding customer demand and improving segmentation (62 percent) and operational challenges (61 percent) as significant barriers to growth.[16] Which are the obstacles holding your company back?

The following questions are designed to help executives assess their current weaknesses and begin to determine which strengths will be their best assets in a converged environment.

- Are you segmenting your customer base according to their communications needs? Do you design new products and services to meet these needs?
- Is your product development cycle fast enough to respond to changing customer needs?
- Do you understand and monitor profitability at the customer and product levels?
- Do you provide a differentiated customer experience? Do you have a "single view" of your customers and all their interactions with your company?
- Is your IT strategy based on flexible, scalable platforms that can integrate all customer-facing processes across your business?
- Is your network design strategy based on open IP standards that will allow integration of multiple future access technologies?

- Is your business "marketing-led" as opposed to "technology-driven"?
- Does your company's organizational structure reflect your focus on customers and their needs?
- Do different parts of your business (i.e., fixed network business, mobile network business) work together to best serve the needs of your customers?
- Have you integrated common functions (e.g., sales, marketing, billing, customer service, distribution, etc.) across different areas of your business to drive revenue and reduce cost?

Conclusion

For telecom providers, convergence means doing business differently. Gone are the days of going it alone. To provide products and services with the functionality and reliability customers expect, telecom companies must pick where they will play and then team up—leveraging what they do best and looking to partners and third parties to round out their offerings.

Understanding how convergence will impact customers, technology requirements, new service creation and current business models is essential for telecom companies to position themselves to differentiate from the competition. For those companies that are able to streamline their organization internally and tailor services externally to quickly evolving and increasingly distinct customer segments, the era of convergence can be one of profitable growth.
Convergence is moving quickly, leaving the unprepared in its wake.

References

1. IBM analysis.
2. PricewaterhouseCoopers. "Global Entertainment and Media Outlook: 2004-2008." June 2004.
3. Ovum. "Wireless Regional Forecasts." 2004.
4. Ibid.
5. "Profiting from WiFi." Communications Today. July - August 2004.
6. IBM/Economist Intelligence Unit, Global Telecom Executive Online Survey, October 2004.
7. Ibid.
8. Ibid.
9. Ibid.
10. Ibid.
11. IBM/Economist Intelligence Unit, Global Telecom Executive Online Survey, October 2004.
12. "Smart antennas will help cure converged devices' woes." Mobile Handset Analyst. Informa Group. November 9, 2004.
13. Carroll, Michael. "Use of Wireless Technology in Cars Still Restricted to High-End Models." Mobile Handset Analyst. Informa Group. November 9, 2004.
14. IBM/Economist Intelligence Unit, Global Telecom Executive Online Survey, October 2004.
15. IBM/Economist Intelligence Unit, Global Telecom Executive Online Survey, October 2004.
16. IBM/Economist Intelligence Unit, Global Telecom Executive Online Survey, October 2004.

Every Bit Better: Next-Generation Optical Transport Platforms

Jörg-Peter Elbers

Director of Technology, Optical Networks
Ericsson GmbH

Introduction

Optical transport has long been the solution of choice for providing high-capacity transmission and switching capabilities in core and metro networks. Being originally designed for time division multiplex (TDM)–leased lines and voice traffic, the synchronous digital hierarchy (SDH) provides an efficient framework for transport of circuit-switched traffic with VC–12 (lower-order SDH: 2 Mbps) and VC–4 (higher-order SDH: 150 Mbps) virtual container granularities. SDH is also extensively used for transport of packet data, but efficiency of data transport is limited in legacy SDH networks because of the rigid structure of contiguously concatenated virtual containers, which do not match the payload rates well in most cases.

The operations, administration, and maintenance (OA&M) functions, as well as the resilience mechanisms developed in SDH, have since then been the benchmark for any carrier-class transport technology. With the introduction of erbium-doped fiber amplifiers (EDFAs), wavelength division multiplexing (WDM) allows multiple SDH channels to be transported over a single fiber without the need of costly electronic regenerators and leads to an increase in per-channel line rates of 2.5 Gbps, 10 Gbps, and 40 Gbps offering aggregate fiber capacities in excess of 1 Tbps.

Fueled by the increasing Internet protocol (IP) traffic, the rollout of residential broadband and high-speed mobile services, as well as the emerging demand for Ethernet-based private line and local-area network (LAN) services, packet-oriented network traffic is continuously growing. To support the traffic growth, the prevalence of different packet-based services requires a packet-optimized transport layer that maintains the carrier-class OA&M and resilience features that SDH offers. As network operators aim at improving their profitability by introducing new revenue-generating services and managing their capital expenditures (CAPEX) and operational expenditures (OPEX), modular platform solutions are necessary. These reduce costs by an increased integration due to technology/layer convergence, lower power consumption/real estate and provide a simpler installation, operation, and maintenance by extended plug-and-play features. The built-in scalability facilitates a future-proof growth in line with increasing traffic demands. While private and business IP services are important in their own right, managed full- and fractional-wavelength services, virtual private networks (VPNs) on Layer 2 (L2), and the efficient transport of storage-area network (SAN) protocols call for an independent transport infrastructure that is truly multiservice (*Figure 1*).

In the core network, it is generally accepted that IP/multiprotocol label switching (MPLS) will eventually form a common virtual-circuit switching backbone layer, running over an SDH/optical transport hierarchy (OTH) transport layer with integrated WDM. Specialized high-capacity traffic streams will be mapped directly onto core transport with minimum overhead. For high-capacity business services, the access and metro networks will have a traffic aggregation and transport role, connecting customer IP routers to IP service switches at the "edge of the core" as well as providing inter-site L2 VPN functionality and support for other protocol transport. For residential broadband and small and medium-sized enterprises, multiservice access nodes (MSANs) will manage the continuing diversity of "last-mile" technologies (including copper, wireless, and some fiber). The transport layer in the access/metro domain employs SDH/OTH–based multiservice platforms complemented by new carrier-class native Ethernet/MPLS–based transport nodes.

This paper provides an overview on emerging technologies for next-generation multiservice transport networks in the core and metro domain and discusses their value proposition:

Next-generation SDH (NG–SDH) extends the life span of SDH by allowing an efficient transport and aggregation of packet-based data services (e.g., IP, Ethernet, SAN) over legacy and new SDH infrastructure.

The OTH introduces an optical layer multiplexing and payload-independent digital wrapper technology that standardizes switching and interworking at wavelength granularity (2.5, 10, 40 Gbps).

Relying on a generalized multiprotocol label switching (GMPLS) protocol suite, the concept of the automatically

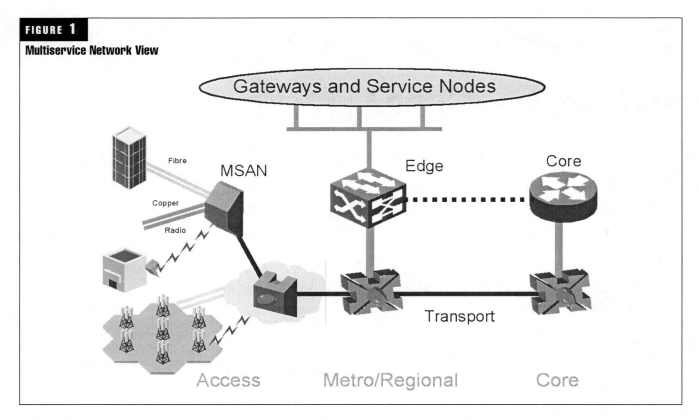

FIGURE 1

Multiservice Network View

Gateways and Service Nodes

MSAN · Edge · Core · Transport

Fibre · Copper · Radio

Access · Metro/Regional · Core

switched transport network (ASTN) provides a topology and connectivity control plane function for transport networks, auto-discovery facilitation, automated circuit provisioning, and mesh-based restoration functions.

Ethernet is extending its coverage from being an enterprise LAN technology into the MAN/WAN segments. As carrier-class Ethernet, it offers virtual line and LAN services using MPLS–based pseudowires or virtual LAN (VLAN)–based provider bridges with OA&M and protection functions comparable to SDH.

Robust transmission techniques simplify 10 Gbps metro/regional transmission by eliminating optical dispersion compensation; self-stabilizing optical rings provide automated optical power control for metro applications. At 40 Gb line rates, advanced modulation formats enable high-speed interconnects on legacy fiber without need for a complex compensation of polarization mode dispersion (PMD).

Reconfigurable optical add/drop multiplexers (ROADMs) and extended optical add/drop multiplexers (EOADMs) are the foundation of a photonic bypass layer that eliminates unnecessary and costly electronic signal regeneration and facilitates an automated provisioning of end-to-end wavelength circuits.

Modular platform architectures relying on a smart integration of optics and electronics help to flexibly configure and optimize a network element for a particular application. With various capacity, switching/aggregation and reach options, they yield lower first-installed costs while preserving headroom for a future capacity buildout.

Network Technologies

NG–SDH

In response to the change from TDM to data-centric transport, NG–SDH was designed to provide a flexible and efficient mapping of data services into SDH along with enhanced capabilities for bandwidth adaptation and network resilience. The mechanisms embraced by the term NG–SDH are the generic framing procedure (GFP), virtual concatenation (VCAT), and the link capacity adjustment scheme (LCAS). As all these extensions need to be implemented at the termination points of an SDH path only, an existing legacy SDH infrastructure can be maintained and past and future investments are protected.

GFP is defined in the International Telecommunication Union Telecommunication Standardization Sector (ITU–T) G.7041 recommendation and describes a standard procedure for mapping packet-oriented or block-coded data protocols into a bit-synchronous channel as offered by SDH/SONET or OTH. It is available in a frame-mapped (GFP–F) and transparent (GFP–T) version: GFP–F works with variable-length frames and provides a one-to-one mapping of data frames into GFP frames. It is optimized for transport of packet-based protocols such as native point-to-point protocol (PPP), IP, MPLS, and Ethernet. GFP–T works with fixed-length frames to encapsulate client characters. As it does not need to wait for full packets or frames to be transmitted, it is optimized for delay-sensitive protocols such as fiber channel, enterprise systems connectivity (ESCON), and fiber connection (FICON). GFP also provides management, performance monitoring, and an optional multiplexing functionality.

VCAT (ITU–T G.707 and G.783 recommendations) defines a method for virtually concatenating SDH connections for transport of traffic exceeding the basic bandwidth of individual lower- and higher-order SDH virtual containers. By bundling an arbitrary number of virtual containers to form a virtual container group (VCG), VCAT solves the problem of the SDH transport bandwidth not matching the payload requirements of data services. For instance, VCAT allows a gigabit Ethernet signal to be carried in seven virtually concatenated VC–4's (VC–4-7v with 1 Gbps payload rate) instead of consuming a whole VC–4-16c (2.4 Gbps contiguous payload rate) as required for transport in an STM–16 signal. The VCs forming a VCG can take different and possibly disjointed paths through the network. VCAT also opens up the possibility to offer fractional data services, where the service interface is, for example, fast Ethernet (FE) or gigabit Ethernet (GE), but the effective traffic rate is only a fraction of this and transport bandwidth, i.e., the number of the VCs in the VCG, is adjusted accordingly (see *Figure 2*).

LCAS is specified in the ITU–T recommendation G.7042. It enables the hitless increase or decrease of the capacity of a VCG in an SDH network. While the individual VCs need to be set up by the network management system, LCAS allows already-established VCs to add to and subtract from a VCG dynamically. As the VCs can be routed over disjointed paths over a network, LCAS can be used as a simple resilience mechanism (*Figure 2*): A single fiber break will reduce the throughput of a VCG (here to 50 percent) but not break the traffic completely. Under normal operation, the 50 percent "protection" bandwidth this scheme provides can be used for generating revenue with additional best-effort traffic.

Optical Transport Hierarchy
With the widespread usage of WDM with data rates of 2.5 Gbps, 10 Gbps, and 40 Gbps, the need emerged to define a client-independent optical layer as well as standardized

amplifier and multiplex section OA&M capabilities. Having its origins in an early definition of a digital wrapper, OTH is defined by the ITU–T G.709 standard and addresses transport, multiplexing, and switching at wavelength granularity. Being asynchronous in nature, literally any digital signal can be mapped directly into the optical data unit payload envelope and transported through the network. Providing a similar level of OA&M and protection functions as SDH, OTH delivers "managed" wavelength services over standardized digital interfaces. Tandem connection monitoring (TCM) allows a client-independent channel monitoring not only for the end-to-end paths, but also for intermediate sections across operator and vendor boundaries (e.g., for a carrier's carrier applications). The inclusion of a robust forward error correction (FEC) with coding gains in excess of six decibels with a standardized Reed-Solomon RS (255,239) code or in excess of eight decibels with proprietary enhanced FEC (EFEC) codes allows greater tolerance to system impairments on the transmission link or the use of less expensive optical transmission technologies for the same system reach. FEC diagnostics can be used to check the health state of a link and to detect degradations before they affect the payload traffic. Used in conjunction with GFP, OTH allows an efficient mapping of high-bit-rate data services such as 1 GE and 10 GE into optical-channel payload units (OPUk).

Figure 3 shows the OTH multiplexing hierarchy and the OTH frame structure with optical payload data units (OPUs), optical channel data units (ODUs), optical transport units (OTUs), optical channels (OChs), and optical transport modules (OTMs). It is important to note that OTH addresses both the optical (OChs) and electrical (OPUs, ODUs, OTUs) switching planes and signals. In addition to single channels, it also defines wavelength division multiplex signals (OTMs) with and without an out-of-band overhead signal (OOS) carried in the optical supervisory channel (OSC) of a WDM system.

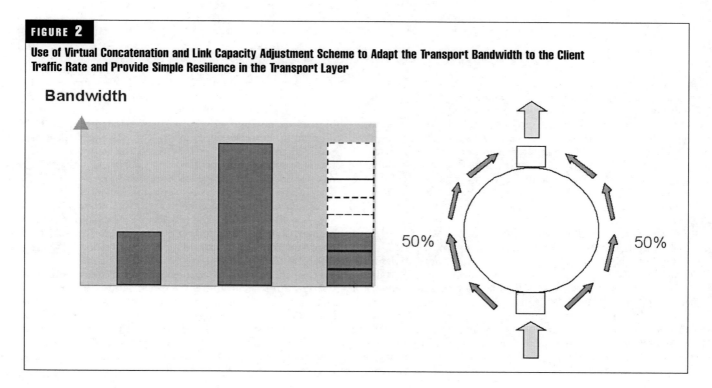

FIGURE 2

Use of Virtual Concatenation and Link Capacity Adjustment Scheme to Adapt the Transport Bandwidth to the Client Traffic Rate and Provide Simple Resilience in the Transport Layer

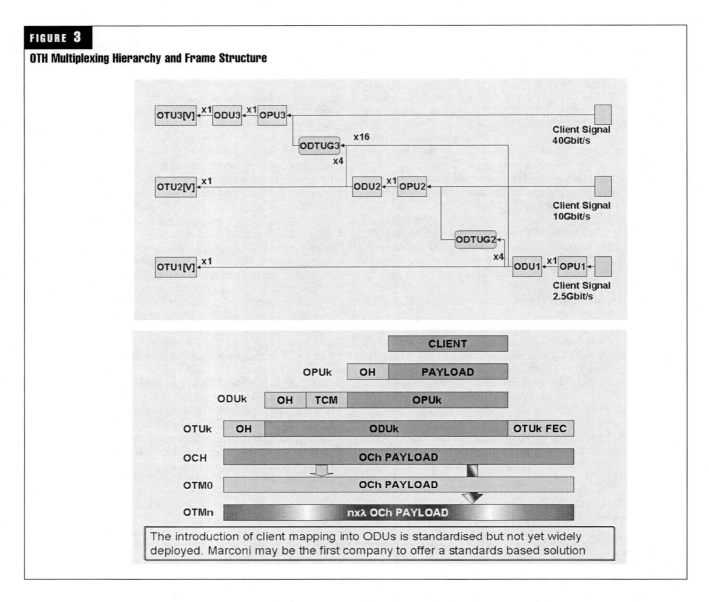

FIGURE 3

OTH Multiplexing Hierarchy and Frame Structure

The introduction of client mapping into ODUs is standardised but not yet widely deployed. Marconi may be the first company to offer a standards based solution

The FEC included in the OTU signals is often sufficient incentive on its own to introduce OTN technology in a carrier network. Modern core cross-connects and multiservice platforms support hybrid VC and ODU switching and are the natural next step to secure revenue from SDH services while migrating to OTH services as the demand for higher bandwidth and clear channel services grows. OTH also offers the possibility to provide managed clear channel wavelength services to large customer premises.

Automatically Switched Transport Network
The automatically switched transport network (ASTN) is standardized by the ITU in the G.807 recommendation and defines topology and connectivity control plane functions for the OTH network. ASTNs aim at off-loading the network management systems by introducing a distributed control plane that uses routing, signaling, and link management protocols to enable functions such as auto-discovery, end-to-end automatic provisioning, and mesh-based restoration for SDH, OTH, and WDM signals. Most commonly employed protocols are the link management protocol (LMP) as well

as the open shortest path first (OSPF) and resource reservation protocol (RSVP) with traffic engineering (TE) extensions. They form part of the GMPLS protocol suite defined by the Internet Engineering Task Force (IETF). Standardized interfaces—both between network nodes (network-to-network interface [NNI]) and between the network and the user equipment (user network interface [UNI])—have been developed by the Optical Internetworking Forum (OIF). They allow interworking between vendors' equipment and across carrier and customer network boundaries. *Figure 4* shows a view of an ASTN–based optical transport network along with the corresponding interfaces either already standardized or being addressed by standardization.

ASTN can be introduced using either a semi-centralized (CD) or a fully distributed (DD) realization. In the centralized version, the routing function resides in a plug-in module of the network management system, while the signaling is implemented as distributed protocol function within the network elements. The CD implementation allows the introduction of pre-planned restoration schemes while preserving compatibility with the traditional network management approach. In the DD realization, both the signaling and the routing functions are distributed. The distributed routing function exchanges topology and resource information between the network elements and allows them to perform their own path selection. The main benefits of the DD implementation are a fast on-the-fly restoration and automatic topology discovery functions. If required by specific deployment scenarios, ASTN CD and DD may coexist in the same network. The network can be partitioned into subnetworks, and these can be connected via external NNI (E–NNI) interfaces. ASTN can also coexist in networks that are partly managed by a traditional network management system. This flexibility enables a gradual introduction of ASTN and a stepwise realization of the resulting cost savings. Customers using ASTN already in live operation calculated 20 percent of CAPEX savings solely from the introduction of ASTN–based meshed restoration.

Carrier-Class Ethernet
As Ethernet is emerging as the preferred universal service interface for leased lines, VPNs and broadband residential access, cost-effective data connectivity services on L2 are of primary interest in the metro and aggregation domain.

Independent of the transport and switching technology used for their realization, different Ethernet service types can be distinguished. Services are classified as point-to-point (line services) and multipoint-to-multipoint (LAN services). The point-to-point can provide a substitute for leased line services earlier realized with PDH/SDH, ATM, X.25, or frame relay. For the multipoint-to-multipoint services the whole transport network effectively behaves like a virtual Ethernet bridge providing inter-site LAN connectivity. *Table 1* shows the terminology defined by the ITU, the Metro Ethernet Forum (MEF) and the IETF. The term "vir-

---

FIGURE 4

Optical Network with Management, Control, and Transport Plane and the Corresponding Interfaces (Source: European IST Project NOBEL)

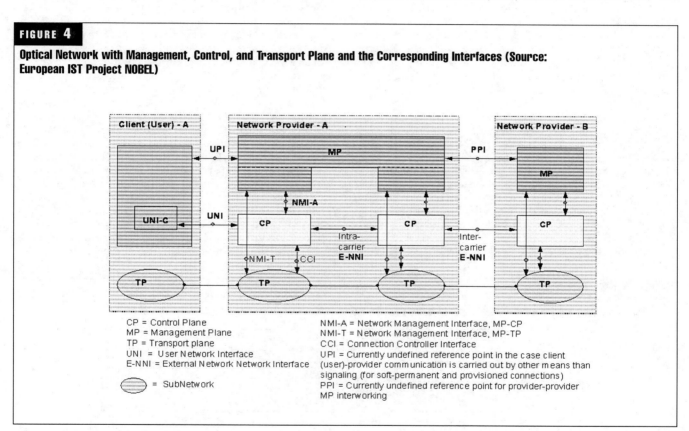

CP = Control Plane
MP = Management Plane
TP = Transport plane
UNI = User Network Interface
E-NNI = External Network Network Interface

= SubNetwork

NMI-A = Network Management Interface, MP-CP
NMI-T = Network Management Interface, MP-TP
CCI = Connection Controller Interface
UPI = Currently undefined reference point in the case client (user)-provider communication is carried out by other means than signaling (for soft-permanent and provisioned connections)
PPI = Currently undefined reference point for provider-provider MP interworking

---

TABLE 1

Classification of L2 VPN (L2VPN) Services

Service type	ITU	MEF	IETF (MPLS pseudo-wire based)
dedicated, point-to-point	Ethernet Private Line (EPL)	Ethernet Line (E-LINE)	Virtual Private Wire Service (VPWS) Virtual Lease Line (VLL)
shared, point-to-point	Ethernet Virtual Private Line (EVPL)		
dedicated, multipoint-to-multipoint	Ethernet Private LAN (EPLAN)	Ethernet LAN (E-LAN)	Virtual Private LAN Service (VPLS)
shared, multipoint-to-multipoint	Ethernet Virtual Private LAN (EVPLAN)		

tual" denotes that the service layer and/or the access links are shared between multiple services. The most prominent technical options for providing these Ethernet services are illustrated by the protocol stacks in *Figure 5*. All protocols can further be encapsulated in OTH/WDM.

Facilitated by the definition of virtual bridging based on VLAN tags in the IEEE 802.1q standard, IEEE 802.1ad provider bridges and DSL Forum WT101 architectures offer alternate ways to transparently switch customer traffic flows in edge and metro networks in a scalable and efficient manner by separating and stacking customer (C–VLAN) and service provider VLAN (S–VLAN) tags. Both methods use the two levels of VLAN tags in a way consistent with the original intended use of VLAN tags, i.e., to partition the traffic into separate broadcast domains. They forward Ethernet frames using the MAC address. The two methods offer alternate ways of mapping the VLAN tags to service type and instance, access equipment, and user ports. Network operators have proposed a number of proprietary schemes that use VLAN tags to forward the Ethernet frames, i.e., the MAC address is not used. Such proprietary schemes require non-standard Ethernet equipment. For simpler configurations, VLAN tagging is not required and packets are forwarded on the MAC address only. The advantage of a L2VPN over a L3VPN for a LAN connectivity service is that the customer edge (CE) and/or access equipment can be a simple Ethernet switch. The provider edge (PE) equipment does not require exchange of routing information with the CE and/or access equipment and does not need to be involved in the customer IP addressing. Alternatively or in addition to the hierarchical partitioning on L2, an interconnection between the provider bridges can be established by means of pseudowire emulation edge-to-edge (PWE3), as defined by the IETF, emulating an Ethernet connection. Based on MPLS label-switched paths (LSPs),

pseudowires on "Layer 2.5" can create a full logical mesh between the MPLS provider nodes over which the switched Ethernet traffic is tunneled. Ethernet and MPLS multicast functionalities can readily be integrated in such a network solution augmenting IP multicast support of IPTV applications. Provider backbone bridges (as being defined in IEEE 802.1ah) and hierarchical VPLS (H–VPLS as being defined by the IETF) are means to further improve the scalability of an L2 provider network by introducing an additional hierarchy level on VLAN tags and MPLS labels. MPLS–based traffic engineering, in conjunction with differentiated services, can be used for providing quality of service (QoS) for different traffic classes.

As OA&M and resilience are critical factors, NG–SDH is a natural candidate for the physical layer of a carrier-class Ethernet solution. For installations in environments where data traffic is predominant, emerging carrier-class Ethernet transport platforms will provide standardized OA&M and resilience on Layers 2 and 2.5 and therefore do not need to rely on proprietary implementations or the existence of an SDH server layer for this purpose anymore. However, if the reliable and accurate transport of timing and synchronization information, as necessary to support legacy TDM services, is required, standardized solutions for Ethernet platforms are not yet available. As SDH was specifically designed for support of demanding TDM services, NG–SDH platforms can best address these requirements. Standardization activities in IETF and ITU are ongoing to address timing and synchronization aspects as well as methods for TDM transport in MPLS– and pseudowire-based networks.

Several standards bodies are defining OA&M and performance monitoring functions at present, amongst them the IEEE, with its 802.1ag standard for Ethernet; the ITU–T, with

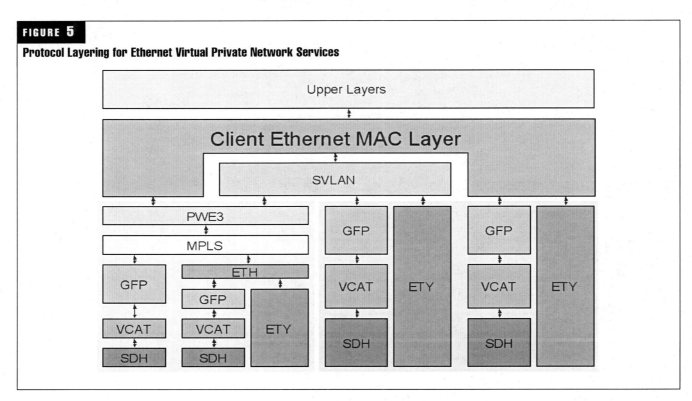

FIGURE 5

Protocol Layering for Ethernet Virtual Private Network Services

its Y.17xx standards that address both Ethernet and MPLS; and the IETF, which is discussing MLPS, L2VPN, VPWS/VPLS, and pseudowire OA&M functions in various contributions. Mechanisms such as the rapid spanning tree protocol (RSTP, 802.1d), multiple spanning tree protocol (MSTP, included in 802.1q), and link aggregation control protocol (LACP, 802.3ad) defined for Ethernet networks by the IEEE, as well as LSP restoration defined for (G)MPLS networks by the IETF, provide some degree of network resilience already. The Ethernet resilience schemes provide recovery in one second or less, i.e., in hundreds of milliseconds. SDH protection mechanisms (50 ms) and MPLS fast re-route (50 ms) can be used to augment the Ethernet mechanisms when the Ethernet mechanisms on their own are insufficient to meet 99.999 percent availability targets. New standardization activities in the ITU and IETF aim at emulating SDH protection switching mechanisms for point-to-point and ring-based topologies on packet, MPLS, and pseudowire level, leveraging the work on OA&M functions for fault detection and indication.

The choice of an NG–SDH or a native carrier-class Ethernet transport platform strongly depends on the network scenario and the infrastructure already available. For the foreseeable future, both solutions have their justification and interworking between them is essential. An SDH–based platform is clearly advantageous where legacy TDM services with stringent timing and synchronization requirements need to be supported.

System Technologies

Robust Optical Transmission

Robust transmission technologies increase high-speed optical signals' tolerance for power fluctuations, optical amplifier noise, and signal distortions and yield an overall simpli-

fied operation as well as an improved signal quality. The improved quality can be translated into increased reach or lower equipment costs. Optical power fluctuations are typically controlled by a fast transient control residing in the optical amplifiers in conjunction with spectral power equalizers for optimizing launch power profiles on a per-channel basis. Amplified recirculating rings (ARR) are becoming commercially available as less expensive solutions for the cost-sensitive metro WDM domain. They facilitate a simple but effective opto-electronic self-stabilization both during normal operation as well as add/drop of channels. FEC and EFEC increases the optical noise tolerance, which, in conjunction with an optimized dispersion map, is also effective in mitigating noiselike impairments as caused by nonlinear inter-channel mixing effects. The noise margin (E)FEC provides allows 10 Gbps channels, for example, to operate at a better performance than 2.5 Gbps channels without FEC.

The use of advanced modulation formats and electronic distortion equalization (EDE) is motivated by different rationales for 10 Gbps and 40 Gbps transmission.

10 Gbps WDM transmission is a well-established technology in volume shipment. As it yields sufficient margins to span several thousand kilometers without electronic regeneration, emerging new solutions need to show cost advantages compared to existing network solutions rather than a further increase in reach. As metro WDM is the major growth area, many of today's new developments are focusing on providing solutions for 10 Gbps metro and regional transmission, which can eliminate optical dispersion compensation over distances of 200 to 300 km and beyond. Cost savings of such configurations can result from the removal of dispersion-compensating fibers as well as from the use of less expensive optical amplifiers. Because of the short link distances, however, achievable savings on a network level

---

FIGURE 6

Example of Combining Inexpensive Transmitter Technologies with Electronic Signal Processing to Obtain 10 Gbps Transmission over 250 km without DCF

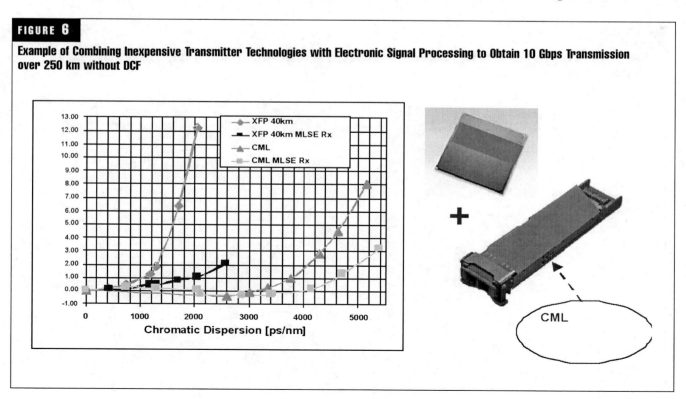

Chromatic Dispersion [ps/nm]

Legend: XFP 40km · XFP 40km MLSE Rx · CML · CML MLSE Rx

CML

are limited, so cost competitiveness compared to standard WDM front ends is also required on an optical interface level. The combination of less expensive transmitters and digital electronic signal processing (the latter just becoming commercially available) is an attractive candidate for such a solution. Combining a maximum likelihood sequence estimator (MLSE) at the receiver side with a 40 km standard transmitter (a chirped managed laser [CMLTM]) allows for 150 (300 km) reach (*Figure 6*). These approaches are particularly interesting as the transmitter/receiver optics can be integrated into hot-pluggable XFP packages, allowing a small footprint and a simplified integration into different transport platforms. With increased electronic processing capabilities, the reach of such a solution can be extended.

40 Gbps is different from 10 Gbps transmission in that 40 Gbps is at the beginning of its product life cycle. Most industry analysts predict a pickup of the 40 Gbps market in the next two years. Main drivers for introducing 40 Gbps services are interconnections between core IP routers. As inter-router traffic grows, the operation of a single 40 Gbps wavelength is easier than of 4 x 10 Gbps. In addition, using 40 Gbps transmission instead of 10 Gbps releases spare capacity for future traffic increase and allows up to four times the transmission capacity compared to today's 10 Gbps systems. As most deployment scenarios plan for a gradual introduction of 40 Gbps, the successful operation over a legacy fiber plant and an already installed 10 Gbps system infrastructure are preconditions. Compatibility with a 50 GHz wavelength grid is a requirement in core networks, whereas in the metro domain 100 GHz channel spacing will be sufficient. The challenge lies in the transmission impairments, which get significantly worse when moving from 10 Gbps to 40 Gbps: for the same reach and modulation format, 40 Gbps requires four times the optical signal-to-noise ratio; the group velocity dispersion tolerance

shrinks by a factor of 16; and the PMD tolerance decreases by a factor of four. Also, optical filtering and nonlinear fiber cause stronger impairments at 40 Gbps than they do at 10 Gbps. As possible improvements by FEC and electronic processing alone are limited for overcoming these impairments, effective 40 Gbps transmission calls for the use of alternative modulation formats. A strong candidate is multilevel phase modulation, which has already shown benefits in various experimental demonstrations. Digital electronic signal processing can further optimize the transmission properties.

Transparent Optical Bypass Nodes

While large-scale optical space switch matrices with thousands of ports were a strategic focus for offering dynamic wavelength services before the Internet bubble burst, the view on transparent optical bypass nodes has now shifted to smaller, more cost-effective solutions. Clearly, a transparent optical bypass can eliminate unnecessary optical-to-electrical-to-optical (O–E–O) conversions, and cost studies have demonstrated resulting CAPEX savings as high as 40 percent for static traffic scenarios. However, such a functionality could in principle be provided by a simple optical patch panel, whereas reconfigurable optical add/drop or cross-connect nodes realize additional OPEX savings from easier network design and the ability to provision new end-to-end services remotely. Main components of modern ROADM and EOADM nodes, set up in a broadcast-and-select architecture, are wavelength blockers and multiport wavelength selective switches (MWSS). Wavelength blockers typically provide power equalization and wavelength blocking on a per-channel basis and are therefore well suited for ROADM nodes. An MWSS forms an WDM aggregate output signal by selectively combining individual wavelengths from its N inputs (see *Figure 7*) and is therefore well suited for EOADM nodes because of its better scalability (cost, performance) with the number of fiber ports. Both blocker-based

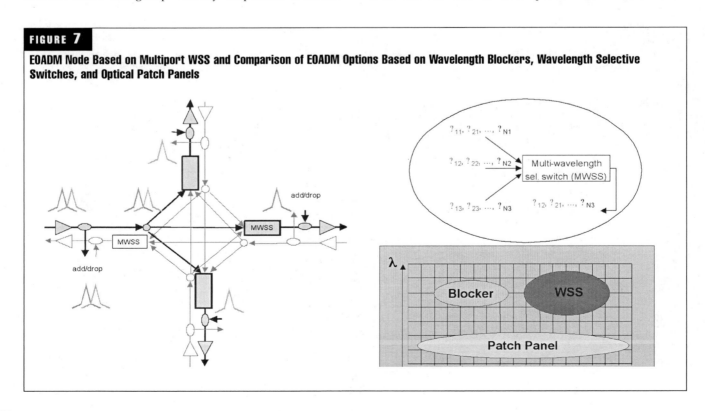

FIGURE 7

EOADM Node Based on Multiport WSS and Comparison of EOADM Options Based on Wavelength Blockers, Wavelength Selective Switches, and Optical Patch Panels

ROADMs and MWSS–based EOADMs are characterized by a high modularity, an integrated power-leveling function, a low cabling effort, low loss, and a broad optical pass band. An important benefit compared to earlier optical cross-connect architectures based on one large space switch matrix is that the broadcast and select architecture does not have a central element that forms a single point of failure. Therefore, no equipment protection is needed and the overall equipment costs are nearly halved. Viewed both from a CAPEX and an OPEX point of view, a static optical patch panel is a good solution for cases where the number of bypass channels is low. With increasing wavelength count, the fiber handling gets more complicated and the cost of the required per-channel power leveling function rapidly reaches that of a blocker or WSS component.

WSS components may also be used to provide a multiport tunable drop filter, i.e., for implementing wavelength agile low-capacity metro WDM nodes. However, this flexibility comes at an additional cost, and based on the network scenario, it needs to be decided whether a fixed filter–, patch panel–, or single blocker–based ROADM node is not a more appropriate option.

Modular Platform Architectures

As highlighted in previous subsections, there is a large range of interface and switching options for realizing next-generation transport platforms. As it is impossible to design specialized node equipment for every singular application, a modular family platform concept with flexible configuration options is essential to address the broadest possible application range and to make best use of optical and electronic functions. For all platforms, plug-and-play and auto-discovery features are indispensable to ease installation, commissioning, and configuration. Additional features are essential to facilitate remote control and fault identification.

On the optical layer, different interface types (SAN, Ethernet, SDH, OTN) and reach options, together with different amplifier and bypass node options, provide multihaul functionality for single-channel and WDM applications. Pluggable optical transceivers on SFP/XFP basis allow even for the use of different interfaces on a multichannel card in the access and metro domain. Future narrowband tunability for SFP/XFP WDM interfaces helps to improve spare part management. Fullband tunable 300 pin MSA modules are standard interfaces for core platforms. Integrated WDM interfaces in electrical switching equipment allow a direct interfacing of the WDM layer without the need of intermediate transponders. Third-party equipment can be connected to the WDM layer via wavelength policy units if interface compatibility can be guaranteed and no electronic demarcation point is required for OA&M and performance monitoring.

On the electronic layer, ODU, SDH, Ethernet, and MPLS switching is provided. Hybrid switch matrices allow a flexible assignment of switching capacity to different switching layers. A multiple-switch matrix approach is used if only a limited amount of switching on one layer is required in a network element that predominantly switches on another layer. Examples of hybrid switches is a combined ODU1/2/3 and VC-4 switch matrix or a protocol-agnostic switch matrix capable of processing SDH and Ethernet/MPLS traffic. Examples of the multiple-switch matrix approach are small Ethernet aggregation switches integrated into optical multiservice platforms or dual data/SDH switch concepts. Increasingly, the functionalities on the electronic layer are becoming more software programmable, allowing customized features and extend equipment functionality to be added over the product lifetime.

Summary

Next-generation optical transport platforms are the foundation of carrier-class multiservice transport networks. Based on a modular concept, these platforms rely on a smart integration of optics and electronics to flexibly configure and optimize a network element for a particular application. Next-generation optical transport is strongly driven by emerging technologies on network and system level: NG–SDH and L2VPN support functions optimize transport networks for the increasing amount of data traffic. OA&M and resilience for circuit-oriented packet-switched networks have SDH–like performance and foster the introduction of carrier-class native Ethernet/MPLS transport. OTH provides a wavelength granular switching layer for optical transport networks with OAM support and FEC for clear channel signals. ASTN brings automatic control plane functionality in core and metro networks for simplified provisioning, meshed restoration schemes, and improved autodiscovery functions. Finally, robust transmission technologies and optical bypass architectures allow a cost optimization of the optical layer with respect to CAPEX and OPEX.

The Necessity of MSANs in the Network Transformation Era

Wu Haining

Director, Fixed Network Marketing
Huawei Technologies Co., Ltd.

Preface

Driven by the forces of market competition and technology advancement, the telecommunications industry is now undergoing a dramatic change that it has never experienced. For one, the prevailing new communication technologies such as mobile communication, IP telephony, and e-mail are nibbling away the fixed-network base. As a result, fixed-network carriers worldwide are seeing their revenue fall and are being forced to transform themselves from a traditional carrier to an integrated information service provider. On the other hand, the accumulation and development of telecom technologies for years has laid a solid ground for service and network transformation, and Internet protocol (IP) technology has matured enough to make it possible for carriers to provide the integrated information services over one converged network. The traditional carriers are now at a crossroads of network transformation. How to successfully carry out the transformation has become the carriers' great concern.

The purpose of network transformation is to support all current and future services over the most economical network and make them sustainable. The new network will not only offer the carriers new opportunities to reap more revenue, but also help to reduce capital expenses (CAPEX) and operational expenses (OPEX).

The Access Network Is Key to the Transformation

The access network is the "nerve end" of the telecom network, through which various telecom services are piped into every home. Therefore, the access network faces a wide range of devices and a complex and intricate environment. As the statistics show, the investment in the access network accounts for more than 50 percent of the money spent in the whole telecom network and the maintenance is as much as 80 percent of the total workload. It suggests that the key point for the network transformation lies in the access network. Also, the current development of the network also proves that the transformation of the access network is vital to that of the whole network.

First of all, the new network must be broadband. Currently, the IP–based data communication has taken the place of

voice traffic as the main traffic of the telecom network, and the demand on bandwidth also continues to escalate. The new network must address the trend of increasing IP–based traffic. While it is easier to make the backbone and transport networks bandwidth- and IP–friendly, we are only beginning to apply broadband to the access network in a small range. The access network once again stands out as the "last-mile" problem, but for bandwidth this time.

Second, the new network must be ready for multiservice delivery and integration. The new network must support the new services derived from the integration of the Internet, mobile network, data network, and telephone network. At present, the various telecom services are partially integrated on the core network and transport network. However, they are carried over independent access networks in the access layer. For example, telephone service is connected via the V5 access network, broadband service via the digital subscriber line access multiplexer (DSLAM) network, and low-speed data service via the defense data network (DDN) access network. While the existence of multiple types of access networks remains a big obstacle to service integration, the large number of network devices keeps CAPEX and OPEX high.

The MSAN Is the Natural Choice

The multiservice access network (MSAN) is an access device burgeoning after the IP DSLAM. MSAN features the strong broadband capability of IP DSLAM and enables the provisioning of multiple services such as plain old telephone service (POTS)/integrated services digital network (ISDN), low-speed time division multiplexing (TDM) data, and frame relay (FR)/asynchronous transfer mode (ATM), as shown in *Figure 1*.

The introduction of MSAN offers a good choice for carriers to transform their access networks. The features of MSAN are as follows:

- *Strong broadband service provisioning capability*: MSAN was born out of IP DSLAM and inherits the strong broadband service-provisioning capability. MSAN supports all kinds of legacy and up-to-date broadband access technologies, including asynchronous digital

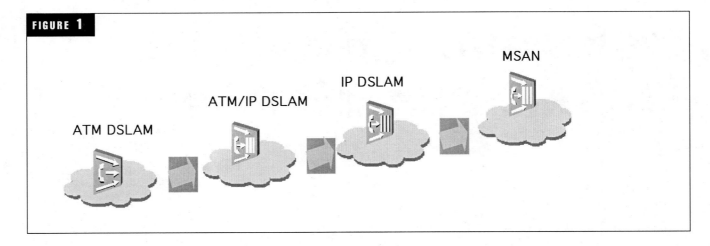

FIGURE 1

subscriber line (ADSL), ADSL2+, single-pair high-bit-rate DSL (SHDSL), very-high-data-rate DSL (VDSL), and VDSL2. It also supports multicast and triple play, traffic aggregation, and fine control and scheduling of bandwidth.

- *Multiple services over a single platform*: MSAN provides the voice over IP (VoIP) capability to adapt POTS/ISDN services to the IP network. It can also work with the core service network to offer a great user experience.

MSAN terminates the low-speed TDM services and adapts the services to the IP network through technologies such as pseudowire emulation edge to edge (PWE3) to implement the integration of the DDN and IP networks. MSAN supports the access of FR/ATM services and bridges the services to the IP network through technologies such as PWE3 to realize the integration of the FR/ATM network and IP network. Therefore, MSAN makes it possible to perfectly integrate the broadband access network, narrowband voice access network, DDN access network, and FR/ATM access network. Such integration helps carriers offer various services through a single platform.

- *Significant reduction in CAPEX and OPEX*: MSAN enables the integration of various access networks to greatly cut CAPEX and OPEX.

○ Reducing CAPEX
  - MSAN reduces the type and number of access devices in the network by integrating DSLAM, V5 access devices (digital loop carrier [DLC]), VoIP access gateways (access media gateways [AMGs]), FR/ATM, and DDN access devices into a single platform.
  - The carriers can reduce the number of the aggregation networks and core networks by constructing the integrated access network with MSAN.

○ Reducing OPEX
  - Fewer devices mean a lower maintenance workload.
  - Fewer devices mean less space occupied and less electrical power consumed.
  - The IP–based integrated platform facilitates service provisioning.

Conclusion

The fixed-network carriers pay great attention to the consolidation and construction of the access network in this network transformation era. As this paper's analysis shows, MSAN is the best choice to realize the network transformation, considering its excellent capability for providing services and cutting CAPEX and OPEX.

Digital Rights Management

Bob Kulakowski
Chief Technology Officer
Verimatrix

DRM Overview

Digital rights management (DRM) is based on technology that enables valuable electronic media content distribution while preserving copyright owners' rights and revenues. DRM associates usage rules or rights for the content, and a well-designed DRM system will use security techniques to preserve the rights while protecting content during distribution. Usage rules can be as simple as restricting copying or allowing only a single copy to be made. Today's media systems and networks require more complicated DRM systems that allow content rights information exchange between devices such as set-top boxes (STBs), personal computers (PCs), and cellular phones.

The goal of a well-designed DRM system is to associate rules for content while protecting the content owner without being burdensome to the end consumer. Rules, often called entitlements, can be associated with a piece of content, attached to the content within the content file, or even embedded within the audio of visual data contents using watermarks. However, DRM rules are attached so the systems must also preserve consumer fair usage rights. It is important to note that a DRM system conveys the usage rights for content and a content security system provides protection of content during the distribution and playback. A DRM system must seamlessly work with a content security system to provide effective security and rights management for the content owners and to provide a simple, easy-to-use system for the consumer.

DRM systems create usage rights that are more enforceable than traditional media security. DRM goes beyond generally understood content protection schemes such as Macrovision that provide analog copy protection or DVD encryption.

Driving Forces

Preserving the rights of content owners and maintaining the economic value of content is the primary goal for DRM. DRM systems are often required by the content owners when a service provider licenses high-value content. A well-designed DRM system may also help prevent service theft, but it will also go far beyond current DRM systems in securing the service. A robust security system combined with a content provider–approved DRM system will increase a service provider's revenue by reducing service and content theft.

Content providers are very astute in their understanding of the use of technology to protect content. As the cost of content production increases, content providers will begin to mandate the implementation of newer techniques to protect content. These techniques will include watermarking and state-of-the-art cryptographic techniques as a prerequisite for content licensing.

The DRM Process

Figure 1 illustrates a simplified diagram of content moving from content owner to service provider. The content files and rights info in the diagram contain the content (i.e., Moving Pictures Expert Group [MPEG] 4 compressed movie file) and rights information in the form of metadata associated with the content.

Rights Information

Content rights information, often called entitlements, describes a consumer's right to receive the content, which is also called rights metadata. The rights metadata for content will include licensing and rights information that the service provider must support in the content distribution chain. Entitlement information, or rights metadata, will be expressed using a form of extensible markup language (XML) for the language syntax. Examples of the information conveyed in the rights metadata include the following:

- The start and end time of the content license
- Whether copies can be made
- Whether the content must be distributed with additional video signal analog or digital copy protection such as Macrovision or digital transmission content protection (DTCP)
- A security system that protects digital interconnections between consumer devices such as DVD players and digital TVs

Entitlement or rights information may also include the number of times the content may be played.

Physical or Encrypted Distribution

Figure 1 illustrates that the distribution of the content and associated rights metadata can be transmitted via a physical distribution channel, for example, when a DVD or digi-beta tape is used to transfer the content from the content licensing agency to the service provider. Electronic distribution is also used when the content is transmitted via an encrypted,

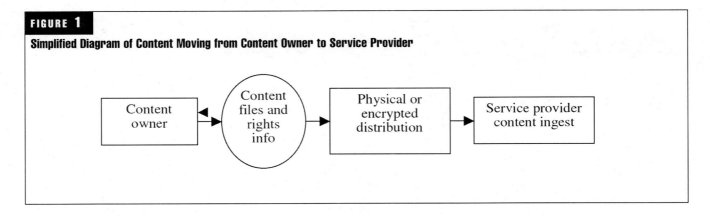

FIGURE 1

Simplified Diagram of Content Moving from Content Owner to Service Provider

secure communications link over the Internet or via a satellite communications system.

Service Provider Ingest
When the content reaches a service provider, there is an ingest process where the content is processed by the service provider's asset management system and the content is entered as a new asset into the system. Re-encryption of the content may occur when this step is performed and all associated rights metadata will be entered into the asset management system and ultimately conveyed to the service provider's clientele. Once the content is entered into the content management system via the on-screen guide or menu system, it will present the asset at the appropriate time to the service provider's clientele.

Content Access by Consumers
When content is accessed by a customer of the service provider, there are several steps that must occur before the content can be played. The service provider's content security system provides content delivery security and will work with the DRM system to convey the content rights to the client player device such as an STB. Credentials of a STB or client device are verified by the content security system in addition to ensuring that the consumer has the appropriate credits. In traditional DVB broadcast systems, this content client credential check is part of the content delivery security and is often performed using information stored in a smart card along with a broadcast of entitlement information to all the STBs in a network. For Internet protocol television (IPTV) systems, security can be established using the two-way network without using a smart card to send entitlements to a single STB and without broadcasting the entitlements to all STBs on a network. In addition, entitlements can be uniquely encrypted so that only one STB on the network can decrypt the entitlement information. Regardless of how the content delivery security system is implemented, the security system must interoperate with the DRM system.

DRM Standards
The DRM system of the future will be based on one or more currently defined or emerging DRM standards. These standards include the following:

- The Open Mobile Alliance (OMA) introduced version 1.0 of their DRM specification designed primarily to protect cellular phone ring tone downloads to consumers. OMA version 1.0 is a simple DRM that

was not adequate for video content, and the OMA has introduced version 2.0, which provides a more robust DRM than that of version 1.0. However, even in the cellular phone industry, the success of OMA version 2.0 has been limited because of a complicated and expensive licensing program and a few competing proprietary DRM systems from Sony, Nokia, and others.

- Microsoft and Real Networks are expanding the scope of their PC–based DRM systems to address more than just PC devices. These DRM systems are important because of their existing subscriber bases.

- Marlin and Coral are two emerging DRM standards that provide complimentary components of a DRM system. Marlin is focused on a single DRM toolkit targeted at client devices that will work with multiple DRM systems. The goal of Marlin is to have a single software development kit (SDK) or toolkit designed to meet the DRM needs of different consumer devices (STBs, PCs, cellular phones, personal digital assistants [PDAs], etc.). Coral is addressing DRM from the standpoint that different services to a consumer may require different content protection systems and, therefore, service interoperability will be necessary. For example, the protection of an audio service may differ from that of a video service. Marlin and Coral are designed to interoperate and provide different components of a DRM solution.

Interoperable DRM
The goal of the interoperable DRM system is to offer consumers persistent DRM and entitlement rights for the content downloaded or purchased. The future of DRM is ultimately driven by the consumer experience.

In addition, the number of issued and pending patents in the DRM space will greatly influence the adoption of a DRM standard. Patent issues and patent infringement concerns may ultimately be the deciding factor in the success of DRM standards. A DRM system must also work with the content security system used to protect a service. The DRM entitlement data must be understood and preserved by the content security system.

Future-Proofing Your Security/DRM System
Because it is too early to know which DRM system will ultimately be the industry standard, the selection of a secu-

rity/DRM system that has been designed to be future-proof is critical. A key future requirement of any DRM system will be to ensure that the interface between the security system and the DRM system is flexible and expandable. Verimatrix has designed a flexible system interface between its security system and any DRM system in the form of a platform independent and system independent DRM interface, which Verimatrix refers to as the entitlement interface.

The requirements of a well-designed entitlement interface must include language, hardware, and DRM independence. Hardware independence means entitlements must be supported and conveyed between hardware devices such as when an STB pushes content to a PDA or PC. Language independence also means that a security system provides interfaces for Java and C++ type programs running under any operating system or even client devices without operating systems. In addition, language independence should entail supporting the rights language metadata of multiple DRM systems, often referred to as the DRM rights expression language. Finally, DRM independence means that the entitlement interface can cross with multiple DRM systems and preserve the rights logic and information supported by a DRM system, that is, when conveying the rights information across a security system to a client device.

Making Sense of Broadband Access

Vineet Kumar

Business Development Manager
Crompton Greaves Limited

Introduction

As competition in the Internet access and telecommunications markets intensifies, service providers find themselves under growing pressure to deliver differentiated, highly competitive services and reduce network costs. In response to this challenge, a powerful class of access devices is increasingly gaining favor and acceptance among small and medium-size enterprises (SMEs) and the providers that serve them.

Integrated access combines voice, data, and Internet access onto a single broadband connection and also lets service providers cost-effectively optimize their local loop infrastructures for the benefit of end users. By consolidating multiple network devices, converging multiple services, and moving intelligence to the network's edge, integrated access lowers requirements for capital equipment, minimizes operational expenditures (OPEX), and maximizes service providers' and carriers' profits. The category of products also enables customers to buy integrated solutions without having to manage their own networks and positions service providers as a partner for providing a wide range of services. By using integrated access to merge legacy networks with evolving infrastructures, service providers can now also enable budget-constrained customers to leverage the power of wide-area networks (WAN) for competitive advantage. In particular, these new services allow SMEs, which often lack the resources to install and manage multiple communications devices, to compete effectively with their larger counterparts in the global marketplace.

What Are Integrated Access Devices?

Integrated access devices (IADs) are compact, scalable access platforms that combine multiple network functions into a single device to lower costs and streamline operations. The same platforms also aggregate voice, data, and access to the Internet and use one broadband connection to replace multiple access lines, each dedicated to a different service.

IADs present service providers with a cost-effective solution that quickly provisions integrated voice, data, and Internet services to all their serving areas without forcing the carriers to overhaul or multiply their network infrastructures. Integrated access processes and routes multiple types of traffic from customer sites to a wide range of carrier serv-

ices, including the public switched telephone network (PSTN), dedicated transmission services such as fractional T1 (Nx64kbps), leased lines and data services such as frame relay and Internet protocol (IP). Also, networks migrate to new standards; the devices support next-generation transmission protocols such as packet-switched asynchronous transfer mode (ATM) and IP. What's more, an emerging enhancement to IADs permits them to be remotely reconfigured through software keys. The resulting ability to automate the upgrades of installed devices gives carriers unprecedented flexibility in deploying broadband solutions and creates substantial opportunities for slashing capital and operational spending.

An IAD that supports time division multiplexing (TDM) and voice-over-packet (VoP) technologies gives service providers the best opportunity to broaden their service portfolios without immediately having to replace existing circuit-switched networks with their packet-based successors. Instead, service providers can deploy IADs in their current local loop infrastructures and continue to serve users with stable TDM–based voice and data services. Then, when carriers are ready, they can merge their networks toward newer services.

IADs for the Local Loop

Deployed at the customer's premises, IADs fall into one of the following three fundamental categories:

Basic Integrated Access Device
Performing the functions of a multiplexer, this type of device resides in legacy networks and exists primarily to aggregate a customer's outgoing voice and data traffic and to channelize onto a single network connection. For voice services, the device will operate as a channel bank and may head an existing private branch exchange (PBX) or key system to aggregate telephone traffic. A foreign exchange station (FXS) interface connects to an analog key system, while a digital system cross-connect (DSX-1) interface attaches to a digital PBX. On the data side, a basic IAD typically provides an Ethernet port and a fractional Nx64k interface for connection into the legacy data network, which can consist of routers, hubs, frame-relay access devices (FRADs), and similar devices.

Enhanced Integrated Access Device
This category of device takes service integration to the next level by providing an all-in-one solution for voice and data.

FIGURE 1

Benefits of Integrated Access

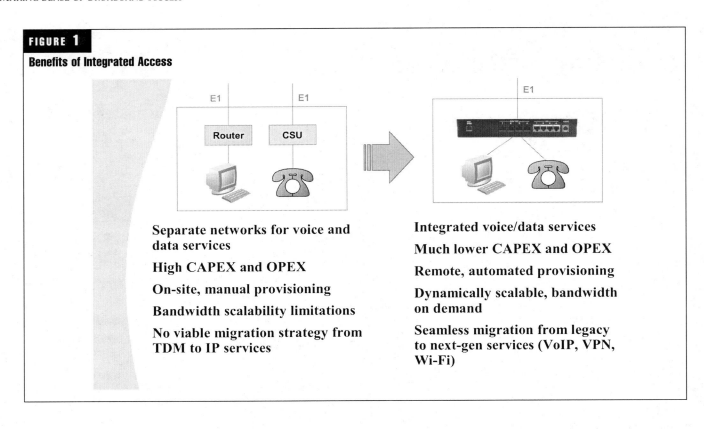

Separate networks for voice and data services

High CAPEX and OPEX

On-site, manual provisioning

Bandwidth scalability limitations

No viable migration strategy from TDM to IP services

Integrated voice/data services

Much lower CAPEX and OPEX

Remote, automated provisioning

Dynamically scalable, bandwidth on demand

Seamless migration from legacy to next-gen services (VoIP, VPN, Wi-Fi)

FIGURE 2

Operators' Network and Service Evolution

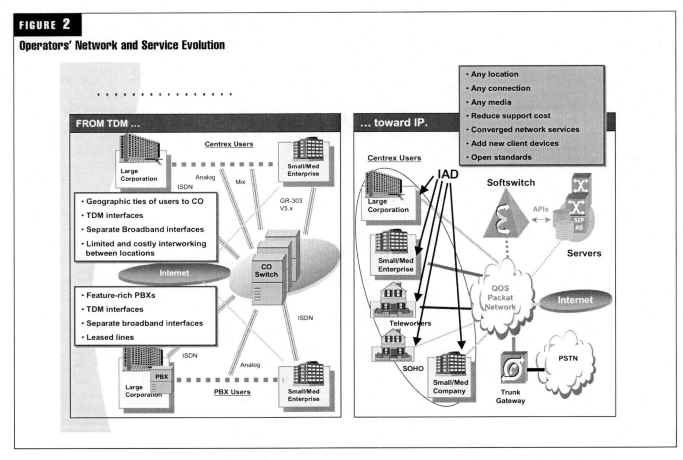

In addition to supporting a channel bank, as basic devices do, an enhanced IAD permits remote management and features integrated routing and Internet firewalling, plus a channel service unit (CSU)/data service unit (DSU), FRAD, and dynamic host configuration protocol (DHCP) functionality. With the same class of integrated platform, service providers also gain the ability to add voice ports or data bandwidth from remote sites. Automated upgrades let carriers lower their operating outlays by eliminating expensive truck rolls, minimize spending for inventory by making one platform adaptable to many configurations, and fine-tune their capacity to customers' requirements. Moreover, enhanced IADs free service providers to pursue new commercial markets; for example, in green-field opportunities, carriers can effectively implement a customer's voice and data network with the help of a single access device. In addition, when IADs further support value-added functions such as local voice switching and localized Centrex services, carriers can provide their subscribers with a complete "office-in-a-box" solution.

Next-Generation Integrated Access

Think of this product classification as a device that resembles enhanced IADs in every respect, including their automated upgrades, except that it also supports new value-added applications and gracefully migrates to voice-over-packet (VoP) services. For carriers that expect to move to new services and packet-based networks, next-generation (NG) IADs provide an ideal answer for the last mile. One of the biggest challenges that providers face when they graduate to new services and protocols is figuring how to make them compatible with—and thus protective of—investments in the existing infrastructure. NG IADs overcomes such difficulties by seamlessly fitting into legacy networks and then permitting them to migrate to new services and packet-based voice and data through downloaded software. Moreover, if NG IADs additionally function as a gateway, they translate all circuit-switched voice and data services to packet-based services and vice versa. Subscribers, therefore, retain all their previous network configurations, including telephone numbers.

IADs for Value-Added Services

Bundled services competitively differentiate carriers from each other and allow them to increase their market share and reduce their customer churn. Service bundles, for example, may include local and long-distance voice in combination with Internet service. What becomes even more compelling is the ability for carriers to distinguish themselves further by offering service packages with value-added functionality. Value-added services provide carriers with incremental revenues without the associated cost of sales or infrastructure. From the perspective of end users, the benefits of value-added services include cost savings, high quality of service (QoS), and increased productivity.

Given the capabilities of integrated access, service providers can effectively offer a portfolio of differentiated services, ranging from a bundle of voice and data services to a fully managed family of value-added services. A value-added voice service may include dedicated Centrex trunk lines for every phone extension in a customer's premise or the ability to deliver local Centrex services that displace a stand-alone phone system. Similarly, value-added data services may include bundled Internet services with e-mail, virtual private networking (VPN), security, and even Web-hosting support.

Managing a feature-rich, value-added service, however, can saddle carriers with operational and support burdens. For a local Centrex service, for example, service providers have to manage and operate all the specific voice features and frequently respond to requests from end users to configure their service. A solution to the challenge lies in giving SMEs access to certain IAD functions that enable subscribers to manage and customize their configurations to include specific features such as the option to turn three-way conference calling on or off. Armed with a capability for flexible management, service providers can own and administer their overall service, simplify day-to-day requirements for configuring end users' services, and minimize the associated operational overhead. *Figure 2* illustrates the process for rolling out value-added services.

Value-Added Services

Although integrated access that offers an all-in-one solution already exists, most qualify as server-based platforms. As such, many of the devices lack some of the key carrier-class attributes that enable IADs to participate in network infrastructures. The characteristics that are necessary for effective participation include the following:

- High reliability
- Resilience (network-equipment building standards)
- Robust, real-time operating system
- Remote management capability that integrates easily with a carrier's high-end operating system

Distributing Intelligence to the Network's Edge

By extending the local loop infrastructure to customer premises, IADs allow carriers to push much of their intelligence for managing services from the core of telecommunications networks to the edge. Localizing service intelligence provides a number of key benefits, including the following:

- Simplified networks and a resulting reduction in accompanying costs
- Increased flexibility in provisioning services and reduced time to market for deploying new services
- Improved ease in expanding networks to keep pace with growth and to enter new markets economically
- Increased reliability through distributed intelligence

To make good on their promised dividends, integrated access must meet the following three critical requirements:

- Local service and service management intelligence
- High reliability
- Centralized management and control of services from network operations centers (NOCs)

Migration to a Next-Generation Network

To compete effectively and to survive, service providers must pursue new market opportunities by creating differentiated services that add significant value. Competition, in

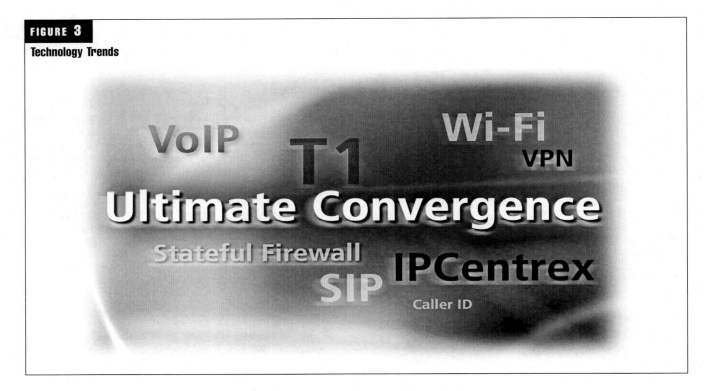

FIGURE 3

Technology Trends

turn, requires carriers to optimize the cost of their networks, shorten time to market and promote the flexible provisioning that will eventually lead to the convergence of voice and data services on next-generation networks (NGNs).

For service providers, however, migrations to NGNs pose many challenges. Carriers cannot afford to scrap their existing infrastructure. Instead, they must somehow adapt their installed networks to support new standards or protocols and protect not only their investments in capital hardware, but also their established revenue streams. Providers, therefore, demand products that dovetail with legacy technology, even as they offer a migration path to packet-based voice and data with minimal operational overhead and impact on service.

By far, the most cost-effective means of migrating to NGNs involves the use of downloaded software, as opposed to truck rolls or forklift upgrades, which both disrupt service and consume lots of scarce resources. Managed remotely from NOCs, the software-based conversion to new technologies is indispensable to a graceful, large-scale migration to NGNs.

Conclusion

An ideal IAD works readily with existing telecommunications assets to meet the needs of end users and service providers alike. In its most basic form, the product category should act as a data multiplexer that integrates a customer's legacy voice and data network. An IAD should also optionally support integrated routing and value-added services. And it should permit software-based migration to packetized voice and data services as carriers see fit.

True integrated access make possible a new level of freedom for carriers that seek to diversify their sources of revenue, hasten the deployment of new services and improve the cost-effectiveness of their maintenance and operations.

Digital Convergence: Who Will Take the Cake?

Gideon Potgieter

Director, Quality and Service
Packard Bell

Executive Summary

In the past, electronic hardware manufacturers were producing goods looking for software; now software content providers are providing content looking for the relevant hardware. Executives of electronic hardware companies need to recognize the risk of content provider companies selecting hardware manufacturers according to their own needs and no longer relying on electronic hardware manufacturers to have their goods available to the public in stores. Personal computer (PC) manufacturers should recognize the risk of substitute products that are emerging and becoming mature, which essentially can perform many of the functions that PCs can perform. They should anticipate the decline of traditional PC usage among consumers and develop a product strategy to be part of the new digital converged world. Content provider or telecommunications companies should see the opportunity of closing the loop and offering the consumer a complete solution, which is a bundle of electronic hardware, telecommunications, and their own content.

When Intel launched its new Viiv strategy, they illustrated the concept of application/service software content providers, PC hardware manufacturers, and consumer electronics (CE) in what they call media devices working together to provide consumers with rich content, including movies, television, music, gaming, and photos. To achieve that goal, Intel proposed that the three industries should work together to achieve such a common goal for consumers.

In today's Western market, consumers are purchasing three distinct product offerings: PCs, CE products such as DVD recorders or MP3 players, and TV and radio signals from content providers via carriers. Consumers have an array of choices in terms of what they can buy through retail stores, but I believe that, increasingly, the content providers will offer packaged bundles, which will include the necessary hardware required by the consumers. This means that traditional electronic manufacturers will have to become contract manufacturers for large content provider companies, either branded or as white goods, whereby the content providers will use their own brand on the electronic hardware. This

also means that consumers will become less and less reliant on traditional PCs to achieve what they set out to do. They will be able to do more and more through the set-top box (STB)[i], including photo editing and viewing, video editing and viewing, music editing and listening, and even creating documents and spreadsheets through technologies that will emerge through the open source community. Software content companies such as Google are playing a big part in this by offering applications that need devices with Internet access, rather than specifying that it has to be a PC.

What Will Happen?

Content providers will become less reliant on electronic hardware manufacturers to have their products available in stores. The impact of this is twofold—CE hardware manufacturers will have to adapt to sell more of their goods to large content providers rather than the classical retail channel, and PC hardware manufacturers will have to have a compelling offer to get consumers to buy PCs rather than fulfilling their computing needs through devices such as STBs.

Who Will Be Affected?

With the exception of those companies that operate in all three segments (see *Figure 1*) such as Sony and Apple, which are PCs, CE devices and software content/service providers—the companies that currently do not operate in all three segments—can be split into two groups: the hardware manufacturers, which again can be split into the CE or PC hardware manufacturers; and the software content/service providers, which could be classical content providers, telecom carriers, or those that have vertically integrated content and telecommunications.

Where Will This Happen?

In early adopter markets such as Japan and the United States, this phenomenon will occur among early adopters who want the latest technology such as digital TV and, more specifically, high-definition TV[ii] (HDTV). HDTV has already started on a large scale in these two countries, both in movie content media (HD–DVD and Blu-Ray discs) and TV broadcasts (as

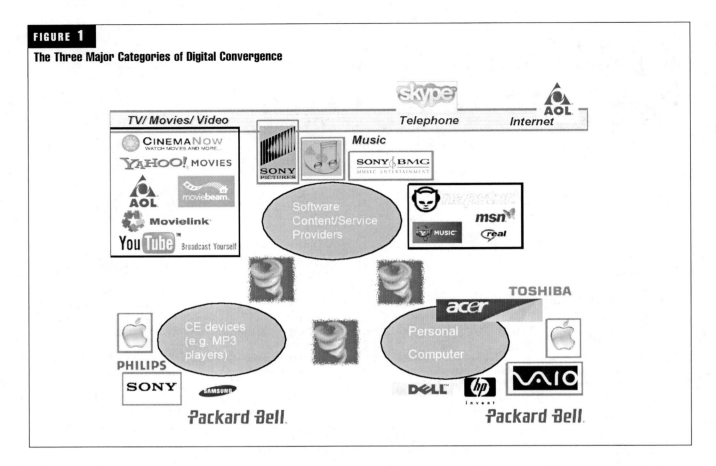

FIGURE 1

The Three Major Categories of Digital Convergence

of September 2005, HD programming is carried by all major television networks in the United States, including ABC, CBS, NBC, and Fox, in at least some broadcast markets).

When Will This Happen?

I believe that, in the early adopter markets, the shift will start taking place over the next two to three years. For the rest of the first-world countries, I believe this shift will start taking place in the next five years, and for countries other than the ones mentioned, in the coming 10 years.

Why Is This Happening?

For early adopters who are interested in entertainment, e.g., movies/TV, music, gaming, or Web browsing, there would be no specific reason to have a discrete PC as an entertainment device, as they will be able to receive the said content through STBs. Some customers might even want time shifting when recording programs, which would mean that those STBs would need hard disk drives (HDDs) inside. PC manufacturers are traditionally strong in working with the way in which software—both the operating system (OS) and the applications—is loaded onto the HDD, so they should have an advantage in how they prepare the HDD for said applications. Many content providers provide on-line storage of significant capacity, which means that if a consumer wants to store some data, it might not necessarily have to be on a local HDD. Software companies such as Google are developing applications that can be done on the Web, regardless of whether one is working on a PC.

How Will It Happen?

As mentioned before, there are three distinct segments at the moment—PCs, CE devices, and software content and service providers. The average consumer in developed countries typically purchases these products and services from various channels. They might purchase their PC from the Web or a specialized retailer, their CE devices from a similar Web store or specialized retailer (but not necessarily the same), and their subscription for television and radio from a separate content provider or directly from a telecom provider. This paper does not aim to cover the convergence of services such as Internet access, telephone, and television (called triple play), nor does it try to cover the question of whether vertical integration will occur between telecom providers and content providers, but rather from the consumer's perspective, that the content providers (whether it be Internet access, telephone, or television) is seen as one dimension of the three main dimensions—PCs and CE devices being the other two.

Threat to PC Manufacturers

When the IBM PC was introduced 25 years ago, early adopters saw the benefits that could be had by performing such functions as word processing and spreadsheets. This was especially true for office users, but later consumers also recognized the benefit of being able to perform these functions at home. This was later followed by multimedia applications such as gaming, photo editing and viewing, video editing and viewing, music mixing and listening, and Internet applications such as surfing the Web and sending and receiving e-mail. Currently, PCs are still mostly used for above-mentioned functions. With the rise of software content/service provider companies such as Google, though, more and more of the said applications can be done through on-line Web applications, through an electronic device (but not necessarily a PC), provided that the said device gives the user access to the Internet. This means that in the future, consumers will no longer need discrete PCs as developed in the original sense over the past 25 years. This is a major threat to PC manufacturers that sell mainly to consumers.

Threat to CE Device Manufacturers

Television content/service providers and other providers specializing in DTV and HDTV that offer their service on their own STBs—which is manufactured by contract manufacturers—means that when consumers make their choice in terms of content, they choose between content providers. Depending on that, the relevant hardware is provided, rather than the hardware of their choice. There are examples of where the service will work on branded hardware that can be bought in retail stores, but this choice is normally limited. Content providers often offer the complete bundle directly to the consumer, which means consumer electronics manufacturers should actively market their goods not only to traditional retailers, but also to the major TV content and service providers that offer special services.

The following are examples of the various categories of CE devices:

Software motion picture content/service providers such as MovieBeam[iii] (Disney) and Movielink[iv] offer their service on their own STB, which is manufactured by contract manufacturers. This means that when consumers make their choice in terms of having content, they choose between content providers, and depending on that, the relevant hardware is provided, not the hardware of their choice.

Software application developer and content provider Microsoft decided to enter the hardware market with their version of a digital audio player, commonly referred to in the market as an MP3 player. Not only will this compete with the market leader Apple, but it will also potentially damage (cannibalize) the hardware sales of other competing hardware manufacturers. This is a perfect example of how a select few hardware manufacturers have the opportunity to license-manufacture on behalf of a company but effectively will no longer be able to compete with their own brand of hardware products.

Napster, a software content provider specializing in digital music files, decided to enter the hardware market with its own MP3 player. This not only effectively sets out to compete with Apple's iTunes (which is the market leader for the distribution of digital music files), but it will also potentially damage the sales of other hardware manufacturers. This is a perfect example of how a select few hardware manufacturers have the opportunity to license-manufacture on behalf of Napster but effectively will no longer be able to compete with their own brand of hardware products.

A lesson can be learned from the mobile telecommunications market. They have managed to provide room for content providers, telecommunications carriers, and hardware manufacturers. Many of the original mobile telephone hardware manufacturers are still commercially active and play an important role in the further development of mobile telecommunications.

What to Do?

PC hardware manufacturers are suffering from a hypercompetitive market, which means they are operating in a very low-margin business. The only way that PC manufacturers can improve their profitability in the short term—apart from the obvious, which is increasing efficiencies and thereby lowering costs—is by bundling services around their hardware products. Some of those services (i.e., extension of warranties, assistance of consumers at home) can be taken care of by the PC manufacturers themselves, while other services (i.e., antivirus programs, Internet access subscriptions, telephony) are provided by software content/service providers. As more and more content/service providers are evolving from single-service offerings into what is commonly referred to in the industry as triple play (data, voice, and video), PC hardware manufacturers need to establish new relationships with the triple-play content/service providers—not only to offer the triple-play content/service providers' services in their bundle (to increase profitability), but also to start offering these content/service providers with their hardware products to establish a reciprocal relationship for the future. Also, the PC hardware manufacturers should strengthen the relationship such that it can become the content providers' contract manufacturer of choice. The also means that PC manufacturers should no longer think in terms of what they can offer in the traditional PC form factor, but rather in terms of innovative new products, which consumers will see as "black boxes," with a graphical user interface (GUI) that look more like what they would expect from a CE device rather than the classical OS GUI.

MP3 audio device hardware manufacturers are by now also suffering from a hypercompetitive market, which means that they are also operating in a very low-margin business. In the short term, MP3 manufacturers need to establish relevant partnerships with software content providers—in this case, digital music providers—so they can bundle the software with their own hardware. Apple has created a complete self-sustaining loop by which an Apple iPod user needs Apple iTunes software and the iTunes store to download music. This means that Apple earns income on the sale of the hardware and repeat sales on the digital music. Although the margins on digital music is razor-thin also because of the high volume (Apple surpassed the 1 billion mark of songs downloaded), the combination of hardware sales and content sales make it very profitable indeed. This business model is similar in a certain sense to game consoles and actual games or PC printers and print cartridges, where hardware manufacturers often accept losses on the hardware only to make their profit on the consumables.

In the longer term, MP3 audio device manufacturers will have to find alternative channels of sales, often called white label, which means they actively need to acquire deals from the software content/service providers that are planning to enter the market for contract-manufacturing those devices, very much like the contract manufacturer for Microsoft's new digital audio player. They need to evaluate the business case of what the potential incremental business could be of contract manufacturing versus losing market share due to more market entrants.

References

IEC: Comprehensive reports: Achieving the Triple Play: Technologies and Business Models for Success

IEC: White Paper: Delivering Economic and Social Development in the Digital Networked Economy

IEC: Comprehensive reports: Delivering the Promise of IPTV

IEC: Comprehensive reports: Broadband Services, Applications, and Networks: Enabling Technologies and Business Models

IEC: Annual Review of Communications, Vol. 58. Digital Convergence: Where Are We Headed?

Intel: www.intel.com/viiv/home.htm

Google: investor.google.com 2006 analyst day presentation, where it shared its strategy and its 70-20-10 framework

Notes

1. An STB is a device that connects to a television and some external source of signal, then turns the signal into content and displays it on the screen. The signal source might be an Ethernet cable (see triple play), a satellite dish, a coaxial cable (see cable television), a telephone line (including DSL connections), or even an ordinary VHF or UHF antenna. Content, in this context, could mean any or all video, audio, Web pages, interactive games, or other possibilities. An STB does not necessarily contain a tuner of its own. A box connected to a television's (or VCR's) SCART connector is fed with the baseband television signal from the tuner and can ask the television to display the returned processed signal instead. This feature is used for decoding pay TV in Europe and was used for decoding teletext before decoders became built-in. The outgoing signal can be of the same nature as the incoming signal, or an RGB component video, or even an "insert" over the original signal, thanks to the "fast switching" feature of SCART. In case of pay TV, this solution avoids the hassle of having a second remote control.

    In the United Kingdom, digital STBs (often called digiboxes, after Sky Digital's trademark for its unit) are usually for digital terrestrial television through services such as Freeview, a service operated by the Freeview Consortium, or through digital satellite with BSkyB and also with digital cable. They are used to access television as well as audio and interactive services through the "red button" promoted by broadcasters such as the BBC with BBCi or Sky with Sky Active.

    Before cable-ready TV sets, STBs were used to receive analog cable TV channels and convert them to one that could be seen on a regular TV.

    Digital STBs are needed to receive digital TV broadcasts because the vast majority of TVs do not yet have such a tuner. In the case of direct broadcast satellite systems such as SES Astra, Dish Network, or DirecTV, STBs are integrated receiver/decoders (IRDs).

    In IPTV networks, STBs are small computers providing two-way communications on an IP network and decoding the video streaming media. Source: en.wikipedia.org/wiki/Set-top_box

2. HDTV refers to the broadcasting of television signals with a significantly higher resolution than traditional formats (i.e., NTSC, SECAM, PAL) allow. Except for early analog formats in Europe and Japan, HDTV is broadcast digitally, and therefore its introduction sometimes coincides with the introduction of DTV. This technology was first introduced in the United States in the 1990s by the Digital HDTV Grand Alliance (grouping together AT&T, General Instrument, MIT, Philips, Sarnoff, Thomson, and Zenith) [1].

    While a number of HDTV standards have been proposed or implemented on a limited basis, the current HDTV standards are defined as 1,080 active interlaced or progressive lines or 720 progressive lines, using a 16:9 aspect ratio in ITU–R BT.709. The term "high-definition" can refer to the resolution specifications themselves or to media capable of similar sharpness such as movie film. Source: en.wikipedia.org/wiki/High-definition_television

3. MovieBeam is a video-on-demand (VoD) service started by Disney. Movies are beamed wirelessly to subscribers' homes using local PBS stations' digital broadcast to deliver the movies to the STB. Up to 10 new movies are delivered to the player each week. The player also contains free movie previews, trailers, and other extras. Source: en.wikipedia.org/wiki/MovieBeam

4. Movielink is a Web-based VoD service offering movies, TV shows, and other material for rental or purchase. The service is owned and operated by Movielink, LLC, a joint venture of Paramount Pictures (owned by Viacom), Sony Pictures Entertainment (including Metro-Goldwyn-Mayer's share when SPE acquired MGM in 2005), Universal Studios (owned by NBC Universal/General Electric) and Warner Bros. Entertainment (owned by Time Warner). Movielink draws its content offerings from the libraries of those studios, as well as Buena Vista Pictures (including Miramax), Twentieth Century Fox, and others on a non-exclusive basis.

    Movies obtained through Movielink can only be viewed on a computer or a TV connected to a computer. However, CE devices such as the Xbox 360 game console will allow users to more easily view these digital media on a traditional TV screen. Experimentation with other business models is under way, including a feature that would allow users to purchase, download, and burn a DVD of a selected film.

    Movielink uses digital rights management software from Microsoft and RealNetworks to protect its content. Consequently, compatibility is limited to Intel-based computers running Microsoft Windows 2000 or XP and a current version of either Windows Media Player or RealPlayer. Also, the service is available only to U.S. residents. Source: en.wikipedia.org/wiki/Movielink

Physical-Layer Challenges in Gigabit Passive Optical Networks

Muneer Zuhdi

Development Manager, Broadband Products
Tellabs

Abstract

The continuously growing popularity of video services and Internet-based applications is fueling demand for higher data rates in access networks. Gigabit passive optical networks (GPONs) are viewed as the next-generation networks that will resolve the major bandwidth bottleneck that currently exists in access networks. However, there are many challenges associated with the design and deployment of these networks.

This paper examines the challenges associated with GPONs and how they can be addressed. The focus will be on the physical layer for the optical data link and video overlay.

Introduction

The strong emergence of the Internet into the public domain has rendered current access networks obsolete. The never-ending need for higher data rates to support new and advanced services has resulted in an increasing pressure on network operators to upgrade their networks. Dial-up and integrated services digital networks (ISDNs) have become a thing of the past. Early broadband deployments based on digital subscriber line (DSL) or cable modem, which use the existing infrastructure to minimize capital expenditures (CAPEX), are anything but future-proof. Since the Internet is a high-growth service, the operators had to invest in their networks to capitalize on this historic opportunity.

Fiber networks have been growing at a much faster pace as compared to telephone and cable networks. Fiber, which has virtually unlimited bandwidth, is more economical than copper in terms of maintenance, upgrade, and manpower requirements. Therefore, the strategic choice for many operators to deliver broadband services is to drive the fiber deep in the access network using a fiber-to-the-premise (FTTP) network.

To minimize CAPEX and operating expenditures (OPEX) of the FTTP network, a passive optical network (PON) with a point-to-multipoint topology will be used to share the OLT with 32 optical network terminals (ONTs) [1]. An FTTP network using PON architecture is shown in *Figure 1*.

Eliminating active electronics in the optical distribution network (ODN) will improve the reliability and maintainability of the network. Furthermore, the point-to-multipoint topology reduces the cost per subscriber, since the central office equipment will be shared among the PON subscribers.

By 2008, global FTTP systems revenues are expected to reach $2.2 billion and PON systems are expected to account for 88 percent of the total FTTP. FTTP subscriptions in North America are expected to grow from 5 percent in 2004 to 23 percent in 2010. [2]

Current FTTP deployment utilizes 622 Mbps downstream data rate on 1,490 nm wavelength and 155 Mbps upstream data rate on 1,310 nm wavelength. The PON is shared among 32 users, and the video service is overlaid on a 1,550 nm wavelength [3]. The wavelength plan specified in International Telecommunication Union (ITU) G983.3 is shown in *Figure 2*.

In order to support Internet protocol television (IPTV) and other video applications (videoconferencing, video chatting, on-line gaming, video cams, peer-to-peer applications, etc.), the data rate will have to be at least four times higher. Hence the need for GPON emerged.

Gigabit Passive Optical Networks

GPON supports a variety of physical line rates. The downstream direction can operate at either 1.2 Gbps or twice that rate, 2.4 Gbps. The upstream direction supports rates of 155 Mbps, 622 Mbps, 1.2 Gbps, and 2.4 Gbps. So GPON can deliver 75 Mbps to each home [4].

There are eight combinations of those downstream and upstream rates. Since the downstream rate must be at least as fast as the upstream, only seven of these combinations are allowed.

The practical GPON deployment that is expected to support all new services will use 2.4 Gbps downstream data rate and 1.2 Gbps upstream data rate. The same 1,490 nm and 1,310 nm wavelength can continue to be used. For many operators, supporting video broadcast through a cable TV–like

FIGURE 1

FTTP Network Using PON Architecture

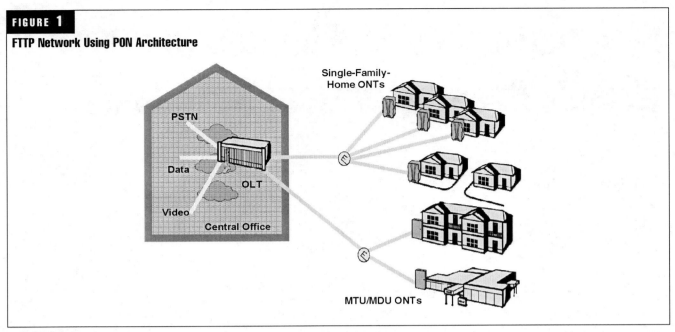

FIGURE 2

Wavelength Allocation

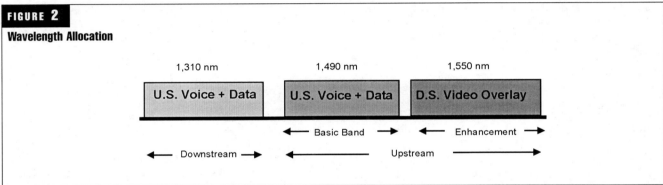

FIGURE 3

FTTP Video Overlay

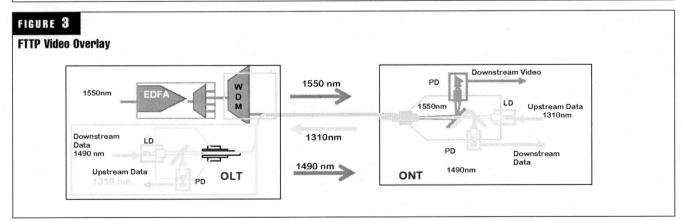

1,550 nm wavelength overlay remains a preferred choice. FTTP with video overlay is shown in *Figure 3*.

As of December 2004, 99 percent of the known video deployment worldwide was through broadcast video and less than 1 percent was through IPTV [5]. By supporting video overlay, operators can market video services immediately with low technological risk while allowing maturation of IPTV middleware and scalability of multicast networks to occur.

Video overlay does not burden the 2.4 Gbps link with the video service that could easily consume 80 percent of the bandwidth. This is particularly true if we consider the recent growth in high-definition television (HDTV) alongside digital (or personal) video recorder (DVR/PVR) technology, and the need for each home to support multiple TVs, DVRs, and computers. Supporting 110 radio frequency (RF) channels with 256–quadrature amplitude modulation (QAM) would translate into more than 4 Gbps, and each home is expected to need more than 50 Mbps [6]. Therefore, carrying video channels using video overlay is considered to be a very efficient use of bandwidth. The return path required for the two-way video communications can still be established over the 1.2 Gbps upstream data link.

Challenges with GPON

There are several challenges that need to be addressed before a GPON network can be deployed. The main challenges are overcoming large loss budgets, mitigating Raman crosstalk, detecting reflection and deterioration in the optical network, and supporting the video network evolution and monitoring. These challenges will be addressed individually.

Loss Budget
In order to have an economical deployment, each PON has to serve at least 32 customers. These customers can be up to 20 km away from the central office. This translates into an ODN loss of 13 to 28 dB. If we add to that the variations of the laser power over temperature, we find that the dynamic range required at the receiver will be challenging to meet using low-cost optical components. An estimate of the ODN losses is shown in *Table 1*.

Optical Components
Supporting the budget in *Table 1* requires high transmit power, low transmit variations, low sensitivity, and high overload. This is particularly challenging considering the data rates that have to be supported and the expected temperature variations. The required optical power levels are listed in *Table 2* [7].

It will not be feasible to support the budget in *Table 1* using conventional p-type–intrinsic–n-type (PIN) diodes. Avalanche photo detector (APD) will have to be used instead to enable the receiver to achieve a sensitivity of –28 dBm. Fabry-Perot (FP) lasers cannot be used either, due to the mode partition noise (MPN), which is exacerbated by dispersion. MPN would result in an inter-symbol interfer-

ence (ISI) penalty [8]. A distributed feedback (DFB) laser will have to be used instead to minimize that penalty.

To extend the loss budget further or to be able to use less expensive optical components, forward error correction (FEC) can also be utilized. The FEC optical coding gain ranges from 3 to 5 dB, and the low incremental cost makes it very economical. So once high-speed PIN–transimpedance amplifiers (TIAs) with less than –25 dBm sensitivity emerge, FEC in the downstream can save at least $10 per ONT by replacing the APD with the lower-cost PIN diode. This directly translates into $10 in savings per subscriber.

Finally, the loss budget can be achieved through power leveling. However, there are many issues associated with power leveling that make it unattractive. The main issue is with interoperability, which is the main driver behind the ITU G983.x for BPON and ITU G984.x for GPON.

Burst-Mode Electronics
The GPON system is designed with efficiency in mind. Overheads for both the downstream and upstream links are shown in *Figures 4* and *5* [9]. The 2.4 Gbps downstream link is 98.3 percent efficient. The 1.4 Gbps upstream link is 88 percent efficient for non-status-reporting dynamic bandwidth allocation, and 93.9 percent efficient for status-reporting dynamic bandwidth allocation. [10]

The low overhead that gave GPON the throughput efficiency has made the design for the burst-mode electronics very challenging.

The laser driver will have to be turned on at all times, which would increase the power requirements at the ONT. Had the overhead requirements been more relaxed, we would have been able to get substantial power savings by turning off the laser driver output stage between bursts. Instead, the ONT will need to use a differential output stage, which burns power even when the transmitter is not bursting.

For the burst mode receiver, the key challenge is related to the consecutive identical digits (CID) requirements. GPON is expected to support CID interval of 72 bits, which is longer than the acquisition time of about 20 ns. This means that the burst mode receiver will need two timing loops and some switching method. A typical implementation is by having the MAC signal the PMD after the end of each burst to enable the fast loop for the next burst. An activity detector triggers the PMD afterward to switch over to the slow loop.

TABLE 2

Optical Power Levels for GPON Class B+

Description	Power levels
Downstream transmit power	+1.5 to +5 dBm
Downstream receive power	–27 to –8 dBm
Upstream transmit power	+0.5 to +5 dBm
Upstream receive power	–28 to –8 dBm

TABLE 1

ODN Loss Budget

	Number of units	dB loss per unit	Total dB loss
1x16 splitter	1	12	12
Fiber (distance in km)	0	0	0
Splices	0	0	0
Connectors	4	0.25	1
Min. loss budget			13
1x32 splitter	1	17	17
Fiber (distance in km)	20	0.38	7.6
Splices	4	0.1	0.4
Connectors	6	0.5	3
Max. loss budget			28

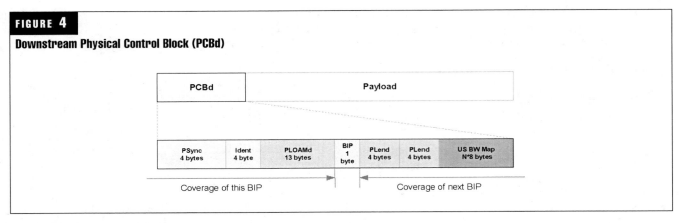

FIGURE 4

Downstream Physical Control Block (PCBd)

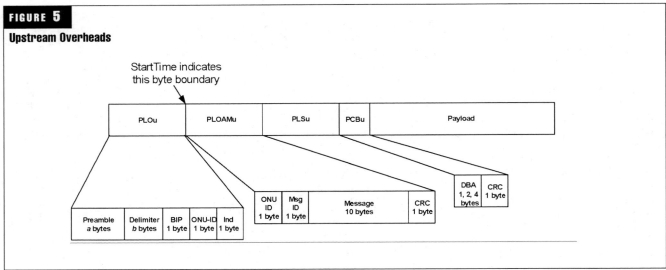

FIGURE 5

Upstream Overheads

Raman Crosstalk

Crosstalk happens when two wavelengths travel on the same fiber, one of them will act as a pump to the other. Nonlinear effects in optical fibers and imperfection in WDM components cause this power transfer that appears as a nonlinear crosstalk. Therefore, part of the power leaks into that wavelength and adds a link penalty that degrades the system performance.

Fiber acts as a Raman amplifier in WDM systems, where longer wavelengths are amplified by shorter wavelengths. The shortest wavelength will be depleted the most, and the stimulated Raman scattering (SRS) will introduce noise in all channels [11]. In the FTTP case, the 1,490 nm wavelength that carries the digital transport will interfere with the 1,550 nm wavelength that carries the video overlay. The result will be a CNR penalty on the video channels.

The 2.4 Gbps signal has limited sensitivity, so the laser has to transmit high power to overcome the large loss budget. This makes the crosstalk worse. However, going to the high data rate of 2.4 Gbps will reduce that crosstalk.

The crosstalk will be most severe in the low video channels, mainly up to 100 MHz. An easy and cost-effective way to mitigate it is by pre-emphasizing the lower four channels by 2 dB to compensate for the 2 dB Raman crosstalk penalty [12]. This will reduce the optical modulation index (OMI)

per channel for the other channels. Since this mainly applies to the lower four channels, the effect will be minimal.

Optical Network Deterioration

When the fiber is driven deep in the network, more challenges arise relative to the troubleshooting and maintenance of the network. This is particularly complex in PONs with the optical splitters in ODN. Dirt, air gaps, and microbends are only some of the conditions that would result in a quick deterioration of the optical link.

The troubleshooting process requires dispatching several technicians to determine if the problem is related to the central office equipment or the fiber network. The technicians will have to be highly skilled and will have to use expensive equipment to troubleshoot the network. This can skyrocket the OPEX and erode the operators' profit margin.

A novel approach to monitor the fiber facility is shown in *Figure 6* [13]. It requires an extra photodiode in the OLT optical block for the detection of the reflected signal. The additional photodiode will have to have very high sensitivity, so an APD is used.

The supervisory elements will monitor the network conditions either in real time or through a maintenance window. The reflection detector will measure the relative signal

FIGURE 6

Block Diagram of the PON Monitoring Elements

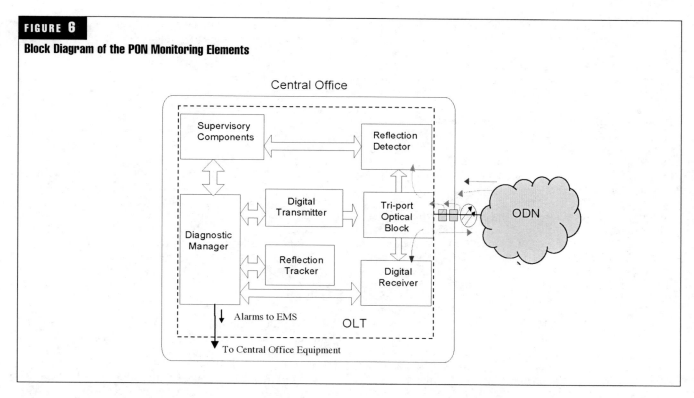

strength of the received reflection. Each relative strength indicator can be translated into an optical received power. At the same time, a counter can be allocated for each reflection. Each counter value can be translated into a round-trip travel time of the optical signal, which in turn can be mapped into a distance. The distance will determine the location of each reflection. The diagnostic manager knows the optical transmit power of the laser, and the fiber attenuation for the 1,490 nm is well known. Thus, we can estimate not only the location of each reflection, but also the approximate reflection value.

Once the reflection value and distance are known, a matrix can be built to track the network condition and alarm for deterioration in the overall optical reflection loss (ORL) performance or for deterioration in the conditions of individual connectors. This matrix is shown in *Figure 7*.

This method provides the operators the necessary diagnostics so a technician can be dispatched only when there is a known problem. The technician will have sufficient details on the problem to resolve it in the shortest possible time.

Video Network Evolution and Monitoring

The video network should be designed to support future upgrades and changes in the channel lineup. Diagnostics need to be added to monitor the 1,550 nm optical input power that carries the video signal along with monitoring the strength of the video signal. Concurrently, the network should be able to remotely adjust the gain of the video receiver. This helps future-proofing the network as the requirements evolve over time.

Several parameters can be implemented to help monitor and troubleshoot the video overlay in GPON. Some of these parameters are as follows:

- *Video lower optical threshold*: To alarm when the 1,550 nm optical input power drops below a predefined threshold

- *Video higher optical threshold*: To alarm when the 1,550nm optical power is higher than a predefined threshold

- *AGC mode*: To specify whether the automatic gain control (AGC) is implemented through feedforward, narrowband feedback, or wideband feedback control loop

- *AGC setting*: To adjust the gain of the video receiver

FIGURE 7

Diagnostics Matrix

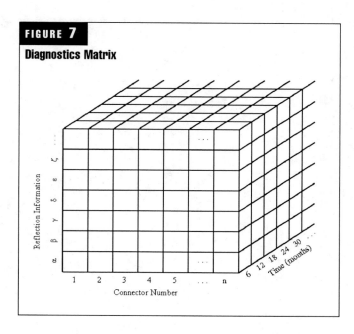

• *Optical signal level*: To monitor the optical level of the 1,550 nm received signal

These parameters along with others provide very valuable information in deploying, troubleshooting, and upgrading PON networks with video overlay.

As the network evolves over time, the operators will have to support less analog channels. By lowering the number of analog channels, the OMI of the digital channels can be increased. This allows the operators to maintain the same quality of video services for a lower sensitivity of the video optical signal.

To illustrate the expected gain in sensitivity when reducing the number of analog channels, a video test setup is shown in *Figure 8*. The setup includes multi-frequency test equipment for the analog channels and a QAM generator and analyzer for the digital channels. The test setup also includes the 1,490 nm and 1,310 nm digital data streams to make sure that any impact from the Raman crosstalk is included. A fiber spool of 20 km is included to simulate the expected distance in field deployment.

The channel lineup that is typically used by cable operators has 82 analog channels and 33 digital channels. The digital channels are 6 dB lower than the analog channels at the input of the transmitter. The total OMI used is 33 percent, which translates into about 3.4 percent OMI for the analog channels and 1.7 percent OMI for the digital channels. The theoretical calculated sensitivity is –6 dBm for a carrier-to-noise ratio (CNR) of 46 dB.

The video receiver limitations for the 82A+33D channel lineup erode the dynamic range in the PON of 15 dB to a mere 7 dB. Therefore, it would be better to minimize the number of analog channels to the absolute minimum required in order to carry local channels along with the public, educational, and government (PEG) access channels. Replacing analog channels with digital channels enables operators to have more programs in the 6 MHz bandwidth and support larger dynamic range due to the less demanding signal-to-noise ratio (SNR) requirements from the digital channels.

First we will look at a channel loading with 40 analog channels, which should be enough to carry local and PEG channels along with 63 256–QAM channels. Using the same assumptions for the receiver, the theoretical video sensitivity is –8 dBm. *Figure 9* shows the empirical data to illustrate how the dynamic range is increased by about 2 dB in the sensitivity.

Another option is to use 30 analog channels along with 63 256–QAM channels. The theoretical video receiver sensitivity is –9 dBm. *Figure 10* shows the empirical data, which shows how the dynamic range is improved by an additional 1 dB.

As mentioned earlier, the ultimate channel lineup has no analog channels. For zero analog channels and 63 digital channels, the theoretical receiver sensitivity is -12 dBm. This brings the overall dynamic range of the 1,550 nm wavelength to about 14 dB, considering that the overload is +2 dBm. The 14 dB range is very close to the 15 dB dynamic range for the data services on the 1,490 nm and 1,310 nm wavelengths. The results of the all-digital-channel loading are shown in *Figure 11*.

As the video network evolves, the ONT video receiver can be accessed remotely to adjust the AGC settings to optimize the gain of the receiver relative to the new channel lineup requirements.

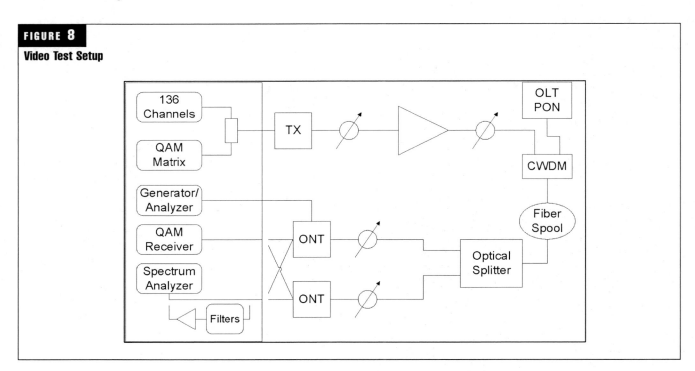

FIGURE 8

Video Test Setup

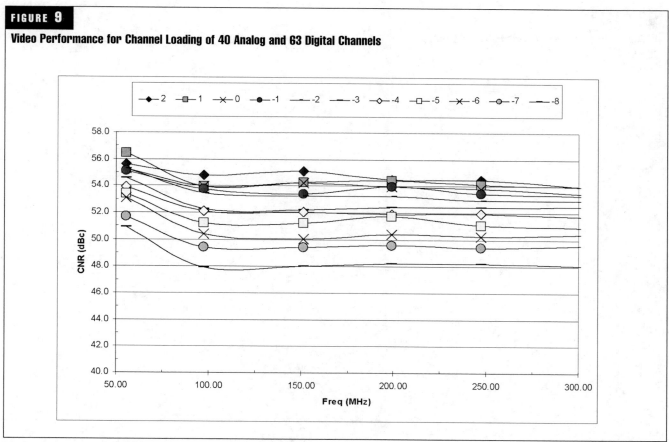

FIGURE 9

Video Performance for Channel Loading of 40 Analog and 63 Digital Channels

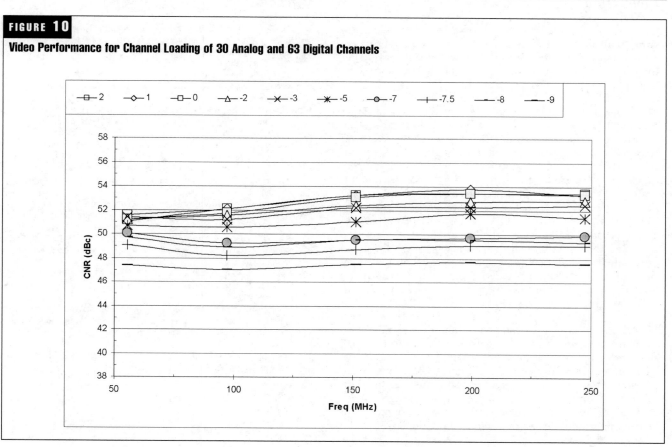

FIGURE 10

Video Performance for Channel Loading of 30 Analog and 63 Digital Channels

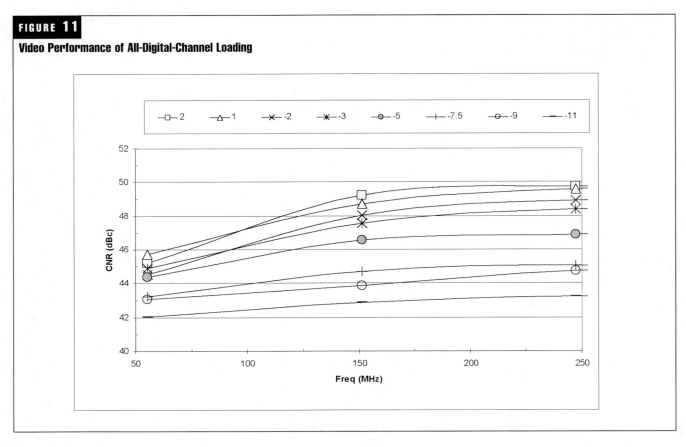

FIGURE 11

Video Performance of All-Digital-Channel Loading

Conclusion

GPONs are the strategic choice for access deployment to support all advanced video and Internet-based applications. The network is future-proof and provides significant savings in terms of future upgrades and maintenance. The throughput efficiency is unmatched.

There are many technical challenges at the physical layer that need to be addressed before moving into large-scale deployment. The throughput efficiency limits the overhead required to ensure error-free transmission at the physical layer. Therefore, the burst-mode electronics will have to quickly recover the data transmitted in the burst while tolerating a long sequence of consecutive identical digits.

The optical components will have to use DFB lasers and APD photodiodes. However, the use of FEC can enable the use of a PIN photodiode at the cost-sensitive ONT. Raman crosstalk from the 1,490 wavelength can be mitigated by pre-emphasizing the lower video channels, and the OLT can be made smarter to continuously monitor the network condition and report any deterioration.

Finally, the ONT can be designed to monitor the video performance and support remote gain adjustment to optimize the performance for different video channel lineups. Migrating to an all-digital-channel lineup helps increase the dynamic range of the video receiver so it matches the digital receiver range. By matching the dynamic range of both the digital and analog receivers, the video deployment will be optimized and the analog receiver limitations will not be a burden anymore.

References

1. ITU–T Recommendation G.983.1, "Broadband Optical Access Systems Based on Passive Optical Networks (PON)," January 2005.
2. T. Mastrangelo, "FTTP Market Outlook Report," The Windsor Oaks Group, December 2005.
3. ITU–T Recommendation G.983.3, "A Broadband Optical Access System with Increased Service Capability by Wavelength Allocation," March 2001.
4. ITU–T Recommendation G.984.1, "Gigabit-Capable Passive Optical Networks (GPON): General Characteristics," March 2003.
5. M. Abrams, "FTTEverywhere," OFC/NFOEC March 2005.
6. D. Piehler, "Video Delivery over the FTTP Network," OFC, February 2004.
7. ITU–T Recommendation G.984.2, "Gigabit-Capable Passive Optical Networks (GPON): Physical Media Dependent (PMD) Layer Specification," March 2003.
8. G. Agrawal, P. Anthony, T. Shen, "Dispersion Penalty for 1.3-um Lightwave Systems with Multimode Semiconductor Lasers," Journal of Lightwave Technology, May 1988.
9. ITU–T Recommendation G.984.3, "Gigabit-Capable Passive Optical Networks (GPON): Transmission Convergence Layer Specification," February 2004.
10. J. Kenny, M. Zuhdi, "A Study of GPON and EPON Efficiencies, QoS, Capabilities, and Physical Layer Characteristics," Tellabs white paper, February 2006.
11. G. Agrawal, Fiber-Optic Communication Systems, Academic Press, 2002.
12. J. Wang, R. Howald, F. Effenberger, M. Aviles, "Mitigation of Raman Crosstalk in BPON Applications," Fiber to the Home Council, September 2004.
13. M. Zuhdi, "Method and Apparatus for Detecting Optical Reflections in Optical Networks," U.S. Patent, January 2006.

Ethernet

An End-to-End Traffic and Hierarchical Traffic Management Architecture for Ethernet-Based Residential Triple-Play and Business Services

David Ginsburg
Vice President, Product Management and Marketing
Riverstone Networks

Xipeng Xiao
Director, Product Management
Riverstone Networks

Richard Foote
Technology Consulting Engineer
Riverstone Networks

Ian Cowburn
Technology Consulting Engineer
Riverstone Networks

Introduction

Quality of service (QoS) from the perspective of the customer is the proper operation of their application and delivery of their services across the network. This is the overall user experience, and it is the user's expectation when purchasing a triple-play or business Ethernet service. The user does not purchase QoS per se, and in most cases could care less about terminology such as delay, jitter, and packet loss. What the user does care about is whether the video, voice, and data applications operate properly between sites or whether the carrier is offering a triple-play service worth paying for.

Factors affecting the user's overall experience include network traffic engineering, hardware and software reliability of the individual network elements, and QoS capabilities of these network elements. QoS also encompasses the following:

- Services delivered to customers connected via legacy technologies such as asynchronous transfer mode (ATM) and frame relay

- Services delivered via digital subscriber line access multiplexer (DSLAM), "active" Ethernet, and passive optical network (PON)

- Mapping of QoS from multiprotocol label switching (MPLS) [1] to virtual local-area networks (VLANs)

- Mapping this to Ethernet

Therefore, a real discussion of QoS requires an understanding of its end-to-end significance, spanning multiple access technologies and network protocols. This paper addresses all of these points.

Another impact on QoS will be whether the network is being operated as a wholesale or retail platform. If the former, the network operator will provide a baseline set of QoS functionality over which third parties provide services. If retail, the network operator is responsible for both the network operation and the services.

Proper support of application QoS requires effective network-level ("macro") engineering and network element ("micro")–based QoS support (*Figure 1*). Both are required, and a major advantage of MPLS is its support for end-to-end traffic protection and engineering. This contrasts with the more basic capabilities offered by VLAN–based Ethernet implementations.

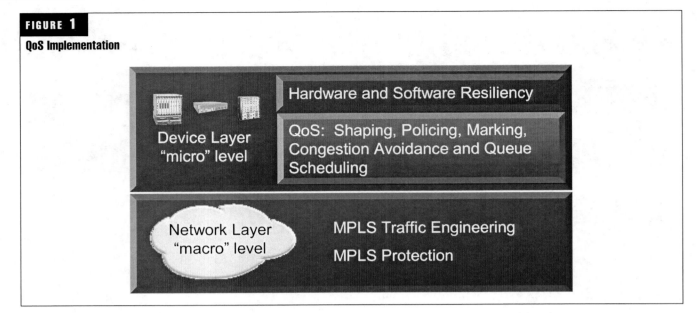

FIGURE 1

QoS Implementation

The Global View

A simple way to look at QoS is from the access network inward toward the network core (*Figure 2*), as follows:

- The customer sends a packet to the network with or without classification and marking (i.e., the customer's gateway, LAN switches, or routers may or may not mark the packet's differentiated services code point [DSCP]). These packets originate on devices sometimes referred to as the customer edge (CE).

- At the edge node within the carrier's access domain, the main classification and marking takes place, possibly taking into account the customer's marking. This classification may be based on the Layer-2 (L2) class of service (CoS) or 802.1p [2], the Layer-3 (L3) type of service (ToS), [3] or DSCP [4]. Marking permits carriers to then differentiate the packets, and may also be based on L2 (CoS), L3 (Internet protocol [IP], ToS, or DSCP), or MPLS EXP. This point in the network is sometimes referred to as the user-provider edge (U–PE).

- Traffic conditioning, consisting of policing (rate limiting) and shaping, now takes place before traffic is forwarded into the carrier's aggregation network. Based on the marking, policing is able to identify traffic out of profile and may either drop it or reset its CoS. Shaping relies on queuing to reshape the traffic flow before forwarding it to the network.

- At the egress interface into the network, congestion avoidance schemes such as weighted random early detection (WRED, described later) will drop packets based on their marking, and congestion management schemes based on queuing will insure that traffic is properly prioritized and scheduled out onto the egress connection. Depending upon the traffic type, both weighted fair queuing (WFQ) and strict priority queuing (SPQ) will come into play.

- The packet is then forwarded to the network core, with its priority remarked as appropriate (i.e., 802.1p to MPLS EXP, described later) when entering the MPLS domain. The device responsible for this is at the network-provider edge (N–PE). Across the core, the MPLS traffic engineering mechanisms come into play at each node crossed.

- At the network egress, the same functions take place.

- Note that in smaller deployments, where MPLS is extended to the edge, the customer provider edge (C–PE) and N–PE functions are combined (i.e., there is no VLAN tier).

Applicable to both the access and core is end-to-end policy management that may involve network admission control. We now look at how these operations take place at the device level.

Network Element-Based QoS

A business or residential service network must be capable of supporting multiple services per customer, each with its own profile. A bandwidth profile can be as simple as the creation of a single best-effort service with some upper limit. However, this model is quickly disappearing, replaced by more sophisticated per-service models. The Metro Ethernet Forum has defined a number of service attributes that a carrier may implement at the network ingress and egress as part of an Ethernet service, and has recently certified networking equipment as being compliant with this definition.

Device-level QoS relies on a combination of traffic marking and policing, shaping, congestion avoidance, and congestion management, as well as system-level reliability that may affect application QoS. This last point is described later.

General Terminology

Across a physical interface, referred to as a user-network interface (UNI) [5], there are multiple Ethernet virtual cir-

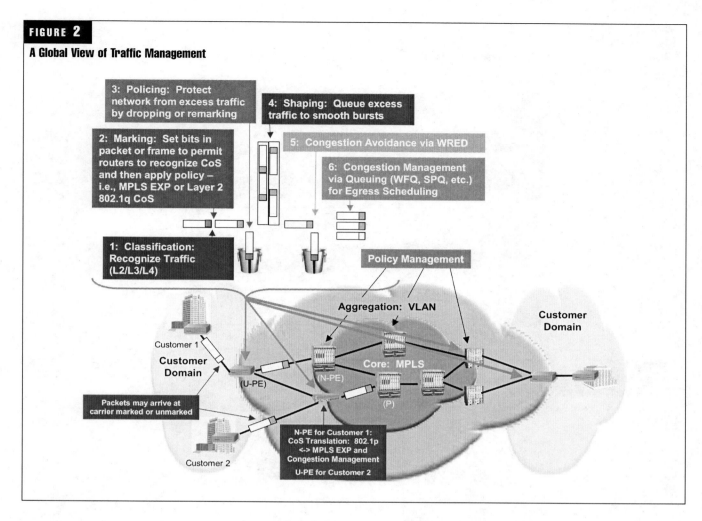

FIGURE 2

A Global View of Traffic Management

3: Policing: Protect network from excess traffic by dropping or remarking

4: Shaping: Queue excess traffic to smooth bursts

2: Marking: Set bits in packet or frame to permit routers to recognize CoS and then apply policy — i.e., MPLS EXP or Layer 2 802.1q CoS

5: Congestion Avoidance via WRED

6: Congestion Management via Queuing (WFQ, SPQ, etc.) for Egress Scheduling

1: Classification: Recognize Traffic (L2/L3/L4)

Policy Management

Aggregation: VLAN

Customer Domain

Customer 1

Customer Domain

(U-PE)

(N-PE)

Core: MPLS

Customer Domain

Packets may arrive at carrier marked or unmarked

(P)

N-PE for Customer 1: CoS Translation: 802.1p <-> MPLS EXP and Congestion Management

U-PE for Customer 2

Customer 2

---

cuits (EVCs) defined (*Figure 3*). And, within an EVC, there may be multiple VLANs or CoSs. Each of these has an ingress bandwidth profile consisting of a committed information rate (CIR), a committed burst size (CBS), an excess information rate (EIR), and an excess burst size (EBS). Commonly, a peak information rate (PIR) is also defined, equal to the CIR plus the EIR. Finally, the maximum burst size (MBS) is equal to the EBS plus the CBS. This is the burst capability of the circuit.

Traffic is marked based on a three-color marker [6] depending upon whether it is: (a) within the CIR (green), (b) above the CIR but below the PIR (yellow), or (c) above the PIR (red) (*Figure 4*). Marking may be based on any of the schemes understood by the network element (i.e., CoS, ToS, etc.) and is also supported when VLAN circuits are used. For review, traffic within the CIR will receive guaranteed bandwidth across the network; traffic above the CIR but below the PIR (and that has not depleted the MBS) will be forwarded if sufficient capacity is available, while traffic above the PIR (and that has depleted the MBS) is dropped. This function is also referred to as policing. Proper network design will ensure that the network is always capable of supporting CIR traffic.

Hierarchical QoS

This basic UNI model is now extended into what is known as hierarchical QoS (H–QoS), a critical element of current

carrier triple-play deployments. H–QoS is just what its name implies—a hierarchical approach of delivering QoS across the network—and contrasts to earlier "flat" QoS. Consider an interface on an Ethernet router. With traditional flat approaches to QoS, traffic arriving at the interface may undergo only a single level of QoS processing. For example, a single customer (many times associated with a single VLAN or local service provider [LSP]) could be defined, with multiple priorities, or multiple customers (VLANs/LSPs) could be defined, each with a single QoS (*Figure 5*).

Although this may be sufficient for more basic business services, and in fact was the order of the day for first-generation metro Ethernet services, it falls down where the carrier is attempting to deliver more sophisticated tiered services to a large number of subscribers. Here, many subscribers may arrive on a single interface, each identified by a VLAN or LSP. Within the VLAN or LSP, the service must support multiple applications (voice, video, and data), each with a different priority. Thus, the requirement for a hierarchical approach to QoS, where the carrier has flexibility in providing per-service QoS, and then an additional level of per-subscriber QoS.

Policing (Rate Limiting)

Rate limiting is usually implemented using a token-bucket algorithm for specific L2/L3/L4 streams of traffic on the

FIGURE 3

Traffic Management at the User-Network Interface

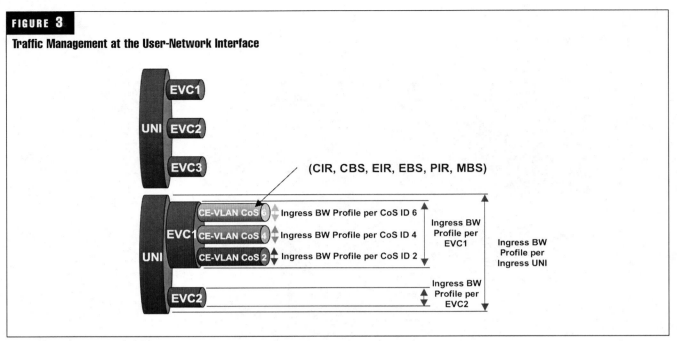

FIGURE 4

CIR/PIR Three-Color Marking

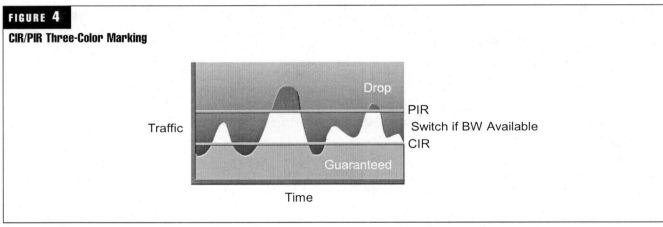

FIGURE 5

"Flat" versus Hierarchical QoS

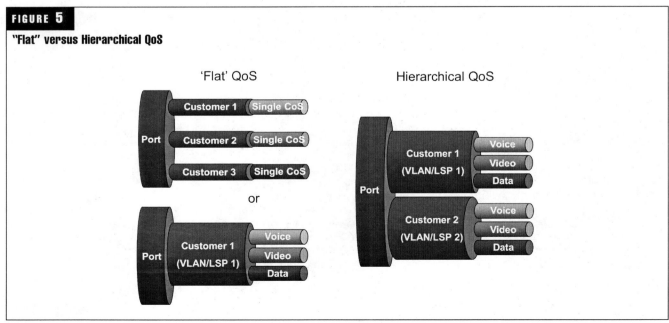

input of a physical interface. Within an Ethernet router, a line card will support some number of limiters in hardware, providing programmable limit + exceed policies. Each policy may be associated with a conform action (rate < CIR) and an exceed action (CIR < rate < PIR). A group of subscriber services (a service access group) may be limited to an overall rate (H–QoS, described earlier). The limiters are also responsible for marking or re-marking the traffic based on the carrier's business model.

Within the network element, rate limiting may operate on the following fields:

- Layer 2—Port, .1Q, or L2 access control list (ACL) (source media access control [SMAC], destination media access control [DMAC], 1p)
- Layer 3—Port or L3 ACL (protocol, SIP, DIP, L4, TOS/DSCP/precedence)
- MPLS: Exp

Actions taken may include the following:

- Layer 2—Drop, transmit, set .1p and transmit, or set MPLS Exp and transmit
- Layer 3—Drop, transmit, set DSCP and transmit, set MPLS Exp and transmit, set precedence and transmit, or set TOS and transmit
- MPLS—Drop, transmit, set .1p and transmit, set DSCP and transmit, set MPLS Exp and transmit, set precedence and transmit, or set TOS and transmit

Rate Shaping

Incoming traffic is normally then passed to a set of shapers. Shaping is similar to rate-limiting in that the identified traffic stream is allowed a specific rate limit and burst size (also based on a token-bucket algorithm). However, traffic shaping also uses buffers to queue packets when excess space is available. When excess queue space is available, traffic shaping allows flows to exceed their assigned rate up to a burst size. This allows traffic to be accepted and buffered when it comes in a burst so that it can be smoothed out. In the shaping stage, WRED is sometimes used to manage congestion.

Shaping may operate on the following classifications:

- Layer 2—Port, .1Q, or L2 ACL (SMAC, DMAC, .1p)
- Layer 3—Port or L3 ACL (protocol, SIP, DIP, L4, TOS/DSCP/precedence)
- MPLS—Exp

Although traffic-shaping queues help keep packets from being dropped, the queues may also add a certain amount of delay, and therefore should be used judiciously with services such as voice over Internet protocol (VoIP).

In a chassis-based system, subscriber traffic is now forwarded across the fabric, maintaining the required eight levels of CoS via SPQ or WRR or hybrid queuing. At the egress, shapers are also required, many times associated with WRED for congestion avoidance when handling transmission control protocol (TCP) traffic (described below). This is important in honoring the CoS under output congestion. At the egress, the overall subscriber data rate is enforced via shapers at the queue and customer level. Port-level traffic

management is enforced through a combination of WFQ and SPQ, ensuring that all customers and traffic types for a customer are prioritized correctly.

Class of Service

Separate from rate limiting or rate shaping, CoS mapping permits marking of an identified stream before forwarding it into the switch fabric.

Classification may be based on the following:

- Layer 2—Port, .1Q, or L2 ACL (SMAC, DMAC, .1p)
- Layer 3—Port or L3 ACL (protocol, SIP, DIP, L4, TOS/DSCP/precedence)
- MPLS—Exp

Congestion Avoidance

WRED is used for buffer management, preventing the buffer (i.e., queue) from becoming full by randomly dropping packets when average buffer utilization exceeds a certain threshold. The reason for doing that is, when a queue becomes full, all arriving packets will be dropped. Such taildrop behavior has a much more severe impact on TCP traffic than random packet drop. WRED is implemented by creating profiles for different traffic types identified via a CoS identifier such as an 802.1p value (for L2 traffic), an IP precedence, ToS, or DSCP (for L3), or an MPLS Exp bit value. *Figure 6* depicts one such WRED profile, with packets set with an 802.1p value of 1 dropped more aggressively (they could be the PIR packets and may occupy less of the queue) than those with a value of 5 (they could be the CIR packets and may occupy more of the queue). In both cases, once the queue depth exceeds the maximum probability threshold, all packets will be dropped. As inferred above, WRED must be applied at any of the contention points within the router and needs to be able to deal with different types of traffic in the same queue and ensure that no TCP–based traffic is excluded from WRED actions.

Congestion Management

Thus far, traffic has been classified, rate-limited, shaped, and re-marked according to the service access group. Incoming traffic is also assigned an internal priority that permits the packets to cross the fabric with a deterministic class of service. In most cases, packets are expected to cross the MPLS network end-to-end with the same priority as assigned per hop.

At the router's egress interface, a scheduling algorithm will select the highest-priority queue first. If packets are waiting, then it is served. If not, then the next highest-priority queue with packets waiting is selected and served. Each queue is assigned a percentage of the total available bandwidth. This bandwidth is always used by packets waiting in the highest-priority queues. However, the remaining bandwidth can be shared by lower-priority queues. As a result, policy queuing enables one to map high-priority traffic such as voice into high-priority queues while other queues share the rest of the available bandwidth. Techniques used at this stage include WFQ and SPQ.

Downstream Considerations

Downstream nodes receiving these packets must ensure they maintain packet ordering. This means that packets arriving with different EXP bits that are related need to enter

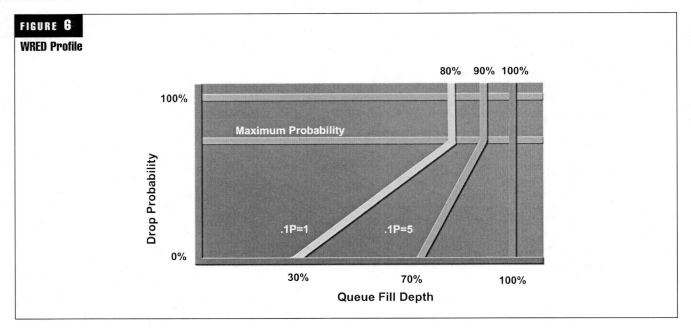

FIGURE 6
WRED Profile

the same queue and be processed with WRED and not individual shapers that could lead to out-of-order packets. These downstream nodes must also perform egress shaping, permitting multiple sites to funnel traffic into a single egress link that is then set to a per-customer CIR and PIR.

Customer QoS

In a commercial model, when limiting and shaping occurs, the customer priorities must be intelligently considered. In these cases, the CIR is filled up to its rate starting with the customer's preferentially marked traffic and continuing with lower-priority traffic until full. The PIR will now fill to its limit with the customer's preferential traffic, and use whatever buffer is left for lower-priority traffic. The customer may choose to use DSCP (here referring to the generic use of the ToS byte) or 802.1p bits to indicate the packet's relative priority. The important point is that it is the customer's choice, not the carrier's, how to mark packets.

In a residential model, the QoS packet markings will be determined by the network equipment between the MPLS PE and the residential in-home network. These elements include the home gateway and any active Ethernet switches, DSLAMs, or PON elements. It is expected that the network will require the same services as the commercial model with one additional requirement—the use of the source IP address and/or the destination IP address for shaping and limiting on the ingress port and egress port.

Other Considerations

The network edge is the proper network element on which this marking should occur for network-wide handling. Each node along the path then must honor the edge markings. In this case, the edge is defined as the edge of an administrative domain. In some networks, it is likely a customer packet will pass through a number of administrative domains, each with their own QoS models and enforcement policies. Regardless of the number of administrative domains the customer packet traverses, the customer packet must remain unchanged.

In cases where the edge of the network is incapable of applying a network-wide QoS marking that meets the requirements of the carrier's model, downstream network elements will be required to enforce the carrier's model.

One problem with rate limiting and shaping is that it relies only on token buckets and not on queues. Therefore, nonconforming traffic may be dropped. For real-time traffic such as VoIP and videoconferencing, where queued traffic may not be useful, this is not a problem. However, for TCP data traffic, this behavior may be sub-optimal. For this reason, at the ingress, rate shaping is becoming increasingly important. Instead of simply marking or dropping the traffic, it is now queued if memory is available.

Customer Connection Scenarios

As part of a triple-play network or a business Ethernet deployment, customers will traditionally connect into the network in one of the following ways:

- They may attach to an edge device that supports VLANs (802.1ad) and may or may not support packet marking. This could be an Ethernet router, switch, or DSLAM.
- They may connect directly to an Ethernet router running MPLS.

Customer Connected via VLANs

The first and most common scenario involves a customer connecting to the carrier network via VLANs (the U–PE scenario in *Figure 2*). Depending upon the sophistication of the carrier's service offering, the customer may connect via one or more VLANs over the single interface, and may also be able to set an application CoS that the carrier will then interpret. These scenarios are commonly deployed as part of a business Ethernet service or residential triple-play based on FTTH, or they may be abstracted to include customers connected to the edge device via Ethernet DSLAMs. We expect that in the future, a greater proportion of business cus-

DAVID GINSBURG, XIPENG XIAO, RICHARD FOOTE, AND IAN COWBURN

tomers will connect directly via MPLS (described in the next section) as technologies such as hierarchical virtual private LAN services (H–VPLS) [7] become more widely deployed and as lower-cost carrier-grade MPLS edge nodes see deployment.

A customer subscribing to a basic VPLS–based intranet service connects to the edge device, and all traffic is mapped into a single VLAN. Here, the edge device ignores any customer CoS markings (this is known as the port-based color-unaware UNI), and the edge-device limits and shapes based on the PIR and CIR. In all models, the .1p is marked accordingly, and then the packet is forwarded across the network. This is then mapped into the MPLS Exp bits with the same values at the VLAN–MPLS boundary.

A more sophisticated model is where the customer maps individual services into different VLANs. Here, the edge device limits and shapes each VLAN individually, and any customer CoS settings are still ignored (known as a port VLAN–based color-unaware UNI). This scenario is commonly referred to as VLAN stacking, since the customer's VLANs are preserved across the carrier's network. At the egress interface into the network, an outer VLAN tag is used to identify the customer, since multiple customers may share the same physical interface.

The next level of sophistication is where the customer has a single VLAN but actually marks the different traffic types (i.e., VoIP and data) with different DSCPs. Here, the ingress of the edge device must properly interpret the markings, and there is a single CoS profile for the port (i.e., different traffic types are still mapped into a single PIR and CIR classification). This model allows the customer to make sure that the carrier is properly handling traffic considered high-priority. This is known as a port-based color-aware UNI.

The final model is where the customer marks different traffic types and maps these into different VLANs. For example, one VLAN could contain only Internet traffic, while the other is a VLAN for voice and data. This latter VLAN contains traffic marked to the different DSCPs as above, and the ingress port properly interprets the customer's CoS markings. This is known as a port VLAN–based color-aware UNI.

Customer Connected via MPLS
Here, the customer connects directly to an Ethernet router operating as an MPLS edge node (also depicted in *Figure 2*). This has the advantage of maintaining the MPLS operational paradigm from end to end, avoiding mapping between various QoS, resiliency, and management schemes. It is also more scalable than an edge VLAN implementation.

The options and UNI models are very close to those in the previous section, with the exception that the mapping to .1p is bypassed—CoS is set directly with the MPLS EXP bits.

End-to-End QoS Examples

Consider residential and business customers connected to services as depicted in *Figure 7*.

The carrier supports VoIP, video on demand (VoD), broadcast video, corporate VPN access, and high-speed Internet access for the residential customer, while the business offering includes videoconferencing and storage backup as well as VoIP and telecommuter VPN access. Across the network, traffic is prioritized based on the mappings in *Table 1*.

Note: The different priorities are many times associated with DSCPs. For example, 6 will map to expedited forwarding (EF), while 4, 3, and 2 will map to different assured forwarding (AF) classes and priorities [8]. Priority level 7 is commonly referred to as network control (NC) and may also map to EF, while 0 is best-effort.

VoIP and video do not require separate CIR and PIR since they are not bursty services, while business data services do. Using terminology common within the industry, the carrier defines 'gold' and 'bronze' data services for the business customer (*Table 2*, below). The gold service is delivered with a CIR and PIR at different priorities, while the bronze service is handled as best-effort.

Multiple customer connections are then aggregated across the uplinks of the DSLAMs, Ethernet routers as fiber-to-the-home (FTTH) serving nodes, or business multiple-tenant units (MTUs). They then arrive at access interfaces of the upstream Ethernet router as members of VLANs, or in the case of some business services, via MPLS LSPs.

Residential QoS Description
The carrier may implement one of two QoS models for the residential customers, and both are described here. The first model maps each subscriber into a separate VLAN, with QoS processed per-subscriber when traffic arrives at the Ethernet router. In the second model, subscriber applica-

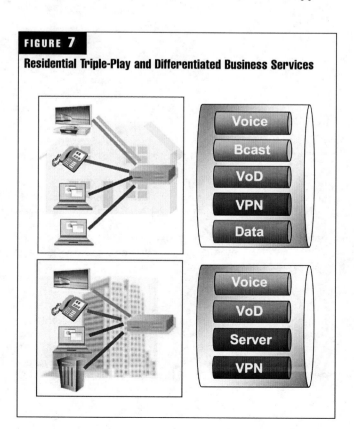

FIGURE 7

Residential Triple-Play and Differentiated Business Services

TABLE 1

Example of Traffic-Class Priorities for Residential and Business Offerings

Priority	Residential traffic type	Business traffic type
7	Network control	Network control
6	VoIP data (bearer) and control	VoIP data (bearer) and control
5		
4	Broadcast video – multicast	Videoconferencing
3	Video on demand	Server-to-server backup gold/green
2		Server-to-server backup gold/yellow
1	Corporate VPN access	
0	High-speed Internet access	Telecommuter VPN access

FIGURE 8

Subscriber-Edge Connectivity

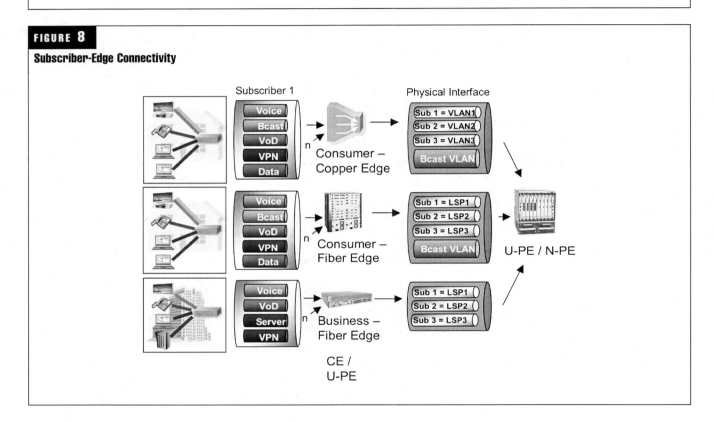

tions are mapped into aggregate per-service VLANs at the DSLAM. Each VLAN is responsible for one application—VoIP, VoD, broadcast video, VPN access, or high-speed Internet access—and QoS is maintained accordingly. Both models are currently deployed as part of triple-play deployments, and both models are effective at maintaining proper end-to-end QoS.

Per-Subscriber VLAN QoS
In the per-subscriber VLAN model, each subscriber is mapped into a separate VLAN at the DSLAM, while the broadcast TV service, carried as multicast, is mapped into a separate VLAN. Within the per-subscriber VLANs, the voice, high-speed data, VPN, and VoD services each receive separate QoS treatment as described earlier.

We first look at the following hypothetical worst-case bandwidth requirement given every service to every subscriber

simultaneously at full encoding rate (for VoIP and VoD) toward the subscriber (DSLAM). In the upstream (ingress) direction, the same traffic engineering is required, but link utilization is much less due to the asymmetric nature of most of the services (data, VoD, multicast).

- The CIR and PIR of VoIP are both 200 Kbps to avoid loss and does not need to burst. All subscribers receive VoIP.

- The CIR and PIR for VoD are both 8 Mbps to avoid loss and also does not need to burst. All subscribers receive VoD.

- The CIR and PIR for the VPN access data service are 1 Mbps and 5 Mbps, respectively. 25% of the subscribers subscribe to the VPN service.

- The PIR for the high-speed data service is 5 Mbps. All subscribers receive this service.

Based on these services, the following holds:

- The aggregated CIR of the subscriber's voice, video, and data traffic for those customers with VPN service is 9.2 Mbps, and the PIR is 17.2 Mbps. The carrier sets a per-subscriber combined-services peak rate (call this a group information rate [GIR]) at 15 Mbps (i.e., though the sum of the peaks is greater, a given subscriber may only receive 15 Mbps of traffic from the network at any given point in time). The VoIP traffic is not counted against this GIR, nor is it subject to GIR buffering (since it is low-latency).

- Assuming 144 subscribers, the carrier would therefore need to allocate .25*144*9.2 = 331 Mbps CIR.

- The aggregated CIR of the subscriber's voice, video, and data traffic for those customers without VPN service is 8.2 Mbps, and the PIR is 13.2 Mbps. The per-subscriber GIR is 10 Mbps.

- Assuming 144 subscribers, the carrier would therefore need to allocate .75*144*8.2 = 885 Mbps CIR.

Based on the sum of the CIRs, the carrier would need to allocate 1.2 Gbps CIR toward the DSLAM (331+885). This is obviously greater than the link bandwidth and would leave no bandwidth for broadcast TV (carried as IP multicast). This multicast traffic adds to the overall bandwidth requirement. In reality, the situation is not that not every subscriber will use every service simultaneously. Note that if they did, and VoD was running at peak utilization, the carrier would easily generate the revenue to bring on-line a second gigabit

Ethernet (GE) interface from the DSLAM if it supported one. This is no different from a carrier engineering business services, tracking link utilization, and bringing on-line additional capacity as required.

In reality, the carrier requires much less bandwidth across the link, calculated as follows and depicted over time in *Figure 9*.

- VoIP at 200 Kbps sustained for two terminals. This is 28.8 Mbps CIR, rounded to 29 Mbps for the remainder of this example, and due to the "minimal" bandwidth requirement (and the fact that folks do tend to talk a lot), assume 100 percent utilization.

- VoD at a mix of 50 percent standard-definition TV (SDTV) (2 Mbps) and 50 percent high-definition TV (HDTV) (8 Mbps). At a given point in time, and at these ratios, 50 percent of the homes view one stream of either type. This is 36*8 + 36*2 = 360 Mbps CIR.

- VPN access for 25 percent of the subscribers is at 1 Mbps sustained, 5 Mbps maximum. This is 36 Mbps CIR and 180 Mbps PIR.

- Internet access for 100 percent of the subscribers at 0 Mbps CIR and 5 Mbps PIR. This is 720 Mbps PIR.

- The "required" CIR for VoIP+VoD+VPN is therefore 29+360+36 = 425 Mbps, a much more manageable number. However, this does not take into account the broadcast traffic.

The broadcast traffic warrants a separate discussion since it has a different traffic profile and is carried in a separate VLAN, and is not included in the per-subscriber CIR of 8.2 or

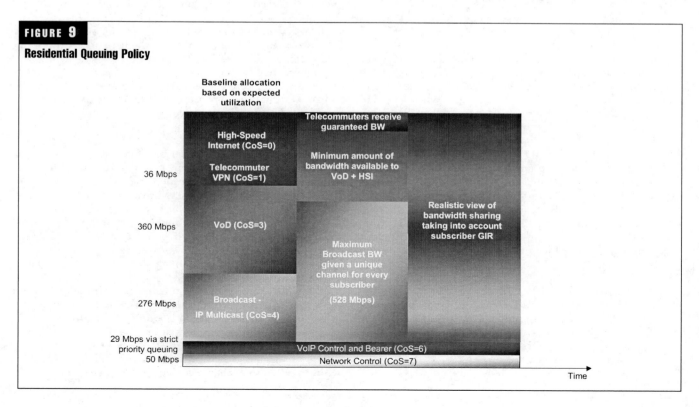

FIGURE 9

Residential Queuing Policy

9.2 Mbps. At the router, there is no 'per-subscriber' identity, since the DSLAM is responsible for multicast replication.

- The carrier transmits 300 IP SDTV multicast channels at 2 Mbps CIR each and 40 IP HDTV multicast channels at 8 Mbps CIR each. The DSLAM maintains state for 30 percent of these channels resulting in 276 Mbps CIR. This is depicted in the first third of *Figure 9*.

- Note that the required CIR for all services is therefore 425+276=701 Mbps, comfortably within the uplink bandwidth, and permitting the data services to exceed their respective CIRs or for the multicast traffic to exceed the planned 30 percent state when required.

- However, under a worst-case scenario, the broadcast traffic could occupy 528 Mbps (40*8 + 104*2). Given that the broadcast TV service defines the service baseline, the carrier must guarantee that the multicast traffic can burst to 528 Mbps if required, even if it will never require this amount of bandwidth (the second slice of *Figure 9*).

- In effect, the "real" CIR of the multicast traffic is 528 Mbps, and the total CIR is now 425+528=953 Mbps, or still within the link capacity.

- Note that some carriers solve this problem by considering 100 percent of the channels in use at the DSLAM, in fact nailing up corner case. With Internet group management protocol (IGMP) fast leaves and joins at the DSLAM, the 30 percent value described earlier is a good working compromise.

- Realistic bandwidth allocation is depicted in the third slice of *Figure 9*.

At the Ethernet router's egress, the VLANs are therefore served in the following way (*Figure 10*):

- Each subscriber is limited to the GIRs as described earlier—10 Mbps for those without VPN service and 15 Mbps for those with VPN service. Within the per-subscriber VLAN, CIRs are served, with the data service(s) occupying the remainder of the GIR.

- The ingress node assigns the VoIP traffic to a rate limiter with the CIR and PIR both set to 200 Kbps so VoIP will not be put into any shaping queue and does not count against the GIR. This minimizes delay and jitter. WRED is not used for VoIP traffic because WRED is for TCP. The node also assigns video and data traffic to an input group shaper, and sets video and VPN data CIR/PIR to 8/8 and 1/5 respectively. Using input group shaping for video or data, the following will take place:

  o All traffic below the CIR will be sent.
  o Traffic between the CIR and PIR will be marked with a certain drop probability. If the total traffic is below is the GIR, the traffic will be sent; otherwise, some packets will be dropped by WRED at the aggregation shaper.
  o All traffic above PIR will be dropped.

- Individual services are now broken out for the next shaping stage. The queues at the egress interface, with each of the services allocated the sum of the CIRs across all subscribers (29 Mbps for VoIP, 360 Mbps for VoD, and 36 Mbps for VPN access).

- Multicast traffic, allocated a CIR of 528 Mbps, is also passed to this second stage. This adds up to the 953 Mbps.

- We will also add 50 Mbps CIR/PIR of network control traffic at this stage, assigned the highest priority.

- Link utilization is now 1,003 Mbps, just over the 1 Gbps link capacity. Note that this assumes optimal utilization. Under small packet sizes, the amount of available bandwidth will decrease, sometimes substantially (i.e., could be as low as 700 Mbps), so this must be taken into account.

- Network control is served first, since it has the highest priority, followed by VoIP. Multicast is next, and although allocated 528 Mbps, in reality it will never require this amount of bandwidth. VoD is served next, followed by VPN and high-speed Internet access.

- Given worst-case multicast, allocated bandwidth is 50+29+528+36 = 643 Mbps, and the remainder is shared by the VoD above CIR, VPN above CIR, and high-speed data.

Now consider the VoD traffic. Below CIR, this is no problem. Conceivably, the VoD could grow to occupy 144*8 = 1.152 Gbps if every home decided to watch an HD movie on-demand some evening. This of course would leave no room for other services. VoD in excess of the CIR will therefore be served fairly, but the bandwidth increment required is a known multiple—it is not variable, as is the case with data. So, for each additional VoD session accepted by the network, there must be a way for the network to inform the router that this bandwidth must be allocated to the VoD pool for the duration of the movie. If the carrier wishes to avoid a situation where video could grow to occupy all bandwidth at the expense of data traffic, a policy management system will be required.

Also remember that the CIRs allocated for multicast and for VoD (as well as for VoIP) are for planning purposes. Although the bandwidth must be available at all times, in reality the content as encoded varies widely from one instant to another. The best evidence of this is with home DVDs, encoded in MPEG2, and with data rates from 2 to 8 Mbps. If one considers a two-hour DVD occupying a single-layer 4.7 GB DVD, the average bit rate is only 5.2 Mbps. Therefore, more bandwidth for additional services is available across the network than may be obvious at first glance.

In the fiber-connected consumer example (middle of *Figure 7*), the same bandwidth considerations still hold, though it is many times easier to add additional bandwidth to the edge node (though possibly not to the downstream BLC if installed). The design exercise above with some changes is also applicable to a PON deployment, taking into account the bandwidth available via a given PON technology.

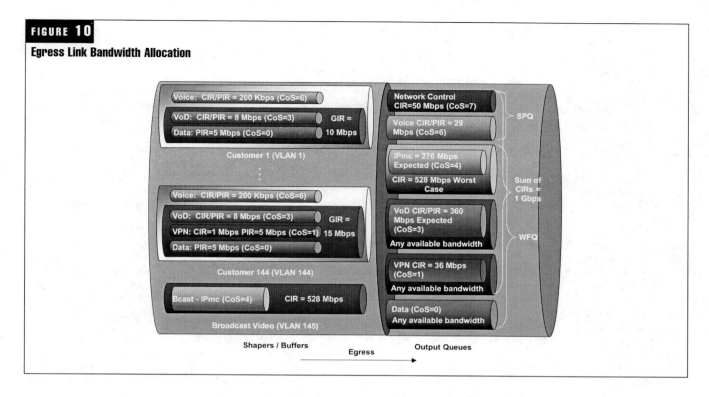

FIGURE 10

Egress Link Bandwidth Allocation

There is in fact an open discussion on prioritization of video traffic. Should broadcast be served at the expense of VoD, or vice versa? Or, should both share a single priority and queue? Some of this will be determined by the service model of the carrier, and they types of IPTV systems deployed. Also note that VoD in this context is true video on demand and not simple point-to-point videoconferencing. This latter application is handled by the VPN or the best-effort data service. As noted above, broadcast is prioritized in this example since it defines the carrier's offering, and the typical customer may not use VoD but will surely view the broadcast channels. For the local exchange carrier (LEC) or municipality to compete against the multiple-system operators (MSOs), the quality of the end-user experience must be equal to or greater than that provided by the MSO. There may only be one customer on a DSLAM viewing the Home Shopping Network at three or four in the afternoon, but that one customer is critical.

Aggregate VLAN QoS
Under the aggregate model, subscriber traffic upon arrival at the DSLAM is aggregated into five per-service VLANs: multicast, VoD, Internet access, VoIP, and VPN access, along with a management VLAN. The multicast VLAN supports broadcast TV and does not result in per-subscriber QoS management at the router, and bandwidth requirements for the management VLAN are minimal. The 144 subscribers handled by the DSLAM share the GE uplink as in the previous example, and all calculations as to the mix of voice, video, and data traffic hold.

Using the priorities described in *Table 1*, VoIP receives the highest priority but requires the least amount of bandwidth. Next is multicast, followed by VoD. However, there must be an upper limit set as to the bandwidth available for this service across the DSLAM's GE uplink, for although the baseline requirement is 338 Mbps, VoD could require more

than a GE if every home wished to view an HD stream simultaneously (144x8 = 1,152 Mbps assuming 1 x HD feed per household). This of course would leave no bandwidth available for other services. Next is telecommuter VPN access, subscribed to by 25 percent of the customers at a sustained rate of 1 Mbps, or 36 Mbps total. Last in priority is the Internet access. This receives the remainder of the available bandwidth, since it is at the lowest priority.

Under a worst-case scenario of multicast, VoIP, VPN, and control, just more than 357 Mbps would be available for VoD (1,000-528-36-29-50); this would be shared with high-speed access (no CIR) and VPN traffic between the CIR and PIR. However, given expected utilization, this worst case will never appear. It is all a factor of traffic and link engineering, and when the carrier notices utilization across the DSLAM uplink that could cause denial of service, revenues will be such that a second uplink should be added.

Given the above considerations, the carrier creates the following queuing policy across the GE access interface (*Figure 8*):

- Network control (CoS = 7) is allocated 5 percent of the available bandwidth, or 50 Mbps.

- VoIP (CoS = 6 and 5) is served with strict priority queuing, and guaranteed 29 Mbps as described above.

- Multicast (CoS = 4) is allocated 276 Mbps based on the profile above but is permitted to burst to 528 Mbps in a worst-case scenario.

- VoD (CoS = 3) is allocated 360 Mbps based on the profile above but may be decreased due to the requirements of the multicast traffic.

- VPN access (CoS = 1) is allocated 36 Mbps. This bandwidth is always available since it is a guaranteed service. It may burst up to 180 Mbps given available bandwidth.

- High-speed Internet access (CoS = 0) is not provided with any guaranteed bandwidth. However, it may use any available bandwidth after the higher priorities are served, including that allocated for VoIP or network control. In a worst-case scenario, it will share the available bandwidth with the VoD service.

At the Ethernet router, subscriber VoIP, VoD, VPN access, and high-speed data traffic is broken out of the per-service VLANs and handled on an individual basis as described earlier.

When traffic arrives at the output line card, VoIP, video, and data traffic will be put into different port queues for transmitting, and WRED may be applied again if needed for video or data.

Network Egress Processing
The last stage in traffic management is forwarding the traffic to the destination routers or servers. Having maintained proper prioritization across the network, traffic of a given type is now forwarded across the proper VLAN or MPLS LSP at the egress interface (*Figure 11*). At the destination Ethernet router, WRED is again applied on the input and output interfaces and traffic is placed in the proper queue.

Business QoS Description

Business customers connected to the network are assigned priorities as depicted in *Table 2*.

The network operator establishes a business queuing policy based on the following parameters across a 10 Gbps interface connected to the carrier's edge device (*Figure 12*):

- Network control (CoS = 7) is allocated .5 percent of the available bandwidth, or 50 Mbps
- VoIP (CoS = 6) is served with strict priority queuing, allocated 50 Mbps
- Video (CoS = 4) is allocated 40 percent of the available bandwidth, or 4 Gbps
- Server-to-server traffic (CoS = 3 and 2) is allocated 20% of the bandwidth, or 2 Gbps
- VPN access (CoS = 0) may use any available bandwidth after the higher priorities are served but is assigned a minimum of 39 percent, or 3.9 Gbps. In *Figure 11*, VPN traffic expands to use bandwidth unused by the video application.

Now consider a single business customer within this 10 Gbps interface. The customer is allocated CIR at 2 Mbps for voice, and CIR at 200 Mbps for video. Server-to-server CoS is set to 3 and the carrier sets a CIR of 50 Mbps and a PIR of 200 Mbps. Input limiting on the 15K is set to these bandwidths for the CIR and PIR, with the CoS reset to 2 for traf-

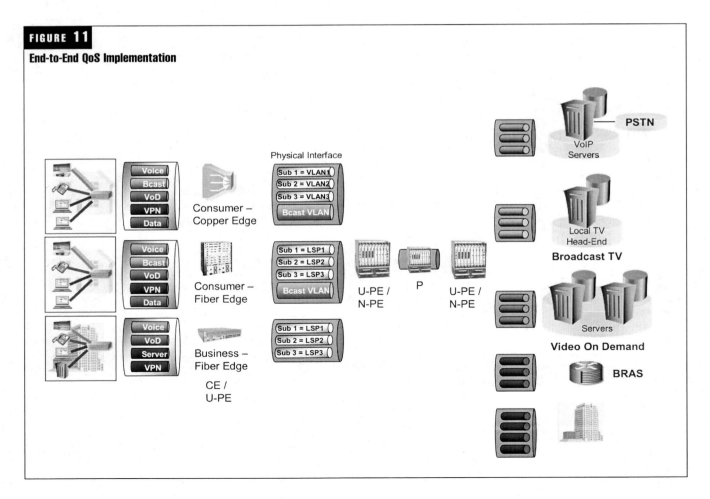

FIGURE 11

End-to-End QoS Implementation

TABLE 2

Mapping Business Traffic to COS/EXP

Priority	COS/EXP	Business traffic type
7	7	Network control (OSPF, RSVP, etc.)
6	6	VoIP data (bearer) and control
5	5	
4	4	Videoconferencing
3	3	Server-to-server backup – gold/green
2	2	Server-to-server backup – gold/yellow
1	1	
0	0	Telecommuter VPN access – bronze

FIGURE 12

Business Queuing Policy

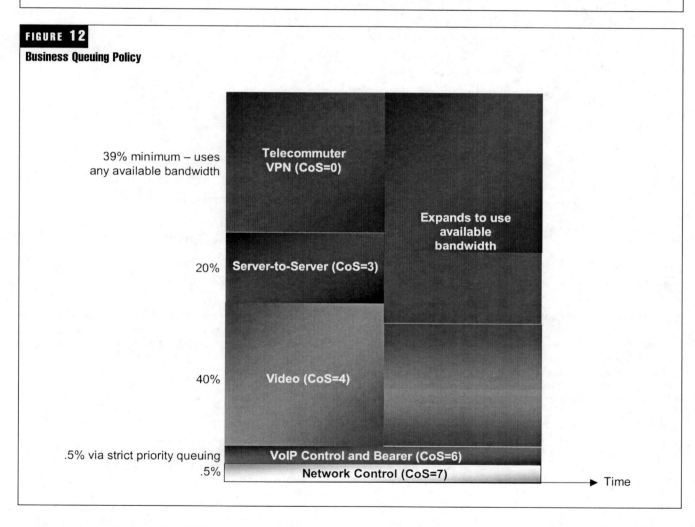

fic exceeding the CIR. Traffic within the CIR remains set to 3. These settings are used at the egress and by downstream nodes for WRED. There is no CIR set for the VPN data traffic. The sum of the CIRs is therefore 252 Mbps. The PIR for data is therefore 1,000-252 or 748 Mbps, so the PIR sum is 748+2+200+200 = 1,150 Mbps. Given that the customer connects via a GE link into the network, the GIR may be therefore set to 1 Gbps.

At the egress, processing needs to take into account the possibility that the output port is congested. This is accom-

plished by setting a WRED profile with a more aggressive drop probability for the yellow traffic. The profile in question will therefore have a more aggressive drop probability for CoS = 2 than for the other CoSs. This profile is then applied at the output physical interface. In addition, the queuing policy created in support of the overall triple-play service is also applied.

System-Level Reliability Impacting QoS
Another factor contributing to the QoS experienced by the end user is the overall network reliability. Given that most

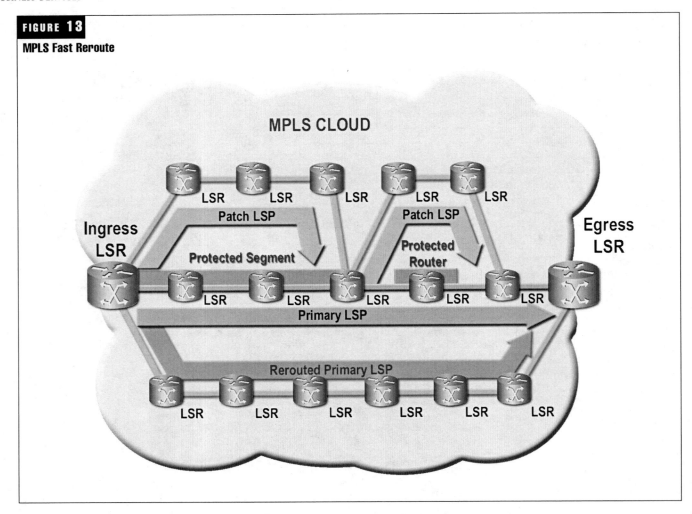

FIGURE 13

MPLS Fast Reroute

network failures are software-driven, the ability of a network element to remain operational across a number of failure scenarios is critical. This forms the basis of the need for a resilient operating system. In fact, according to a recent report from Network Strategy Partners [9], the average downtime per outage with modular software is 1.5 seconds, whereas the average for monolithic software is 25 seconds.

Network-Based QoS

Traffic management at the device level is only one part of the total solution. In order for it to be effective, the network must preserve application QoS end-to-end. This is the role of MPLS.

MPLS Traffic Protection

An important capability of MPLS is network fault protection via the establishment of backup paths. The network operator will configure the MPLS LSP from the ingress router that is fast reroute (FRR)–enabled (*Figure 13*) [10]. When there is a network failure, the router upstream of the protected segment (the protecting router) will detect this via L2 notification. Within 50 to 100 ms of the failure, a "patch" LSP is then used to carry traffic around the protected segment. This failover time prevents noticeable application degradation, but the patch LSP may follow a sub-optimal routing across

the network. Therefore, the protecting router sends a message to the ingress router requesting it to compute a new optimal path. One way of optimizing this process is by creating a hot-standby LSP in advance.

MPLS Traffic Engineering

On par with network protection is the requirement to properly size the network links. Since the network topology and link capacity does not (and should not) change dynamically, uneven traffic distribution may create high link utilization or even congestion in one part of the network, even when the total capacity of the network is greater than the total customer demand. This may be due to shifting traffic demands or random incidents such as fiber cuts or failed transit nodes. Proper traffic engineering relies on the provisioning of separate LSPs for different classes of traffic. For example, from a given ingress router, one LSP will handle premium traffic, with another handling assured and best effort traffic. The operator sets a maximum bandwidth that may be used by premium traffic on its LSP to maintain its QoS, with any available bandwidth used by the other traffic types. Traffic from customers is limited, shaped, and marked at the ingress based on the network operator's selection of one more L2 to L4 fields. This traffic is now forwarded across the proper LSPs.

This is closely related to the concept of per–VC LSP shaping, permitting the carrier to offer a multipoint frame relay–like model while helping to control core bandwidth utilization by avoiding the n*access rate bandwidth consumption. This allows a customer to purchase bandwidth (CIR and PIR) for a hub site and for each of the remote sites. The carrier then can guarantee these rates while conserving core bandwidth.

References

[1]   E. Rosen et al., "Multiprotocol Label Switching Architecture," RFC 3031.

[2]   "ISO/IEC Final CD 15802-3 Information technology – Tele-communications and information exchange between systems – Local and metropolitan area networks – Common specifications – Part 3: Media Access Control (MAC) bridges," Institute of Electrical and Electronics Engineers (IEEE) P802.1D/D15.

[3]   P. Almquist, "Type of Service in the Internet Protocol Suite," RFC 1349.

[4]   K. Nichols et al., "Definition of the Differentiated Services Field (DS Field) in the IPv4 and IPv6 Headers," RFC 2474.

[5]   "MEF 11: User Network Interface (UNI) Requirements and Framework," Metro Ethernet Forum.

[6]   J. Heinanen, R. Guerin, "A Two Rate Three Color Marker," RFC 2698.

[7]   M. Lasserre, V. Kompella, "Virtual Private LAN Services over MPLS," draft-ietf-l2vpn-vpls-ldp-08.txt.

[8]   J. Heinanen et al., "Assured Forwarding PHB Group," RFC 2597.

[9]   "Business Benefits of Modular Software in Carrier Ethernet Routers," Network Strategy Partners, www.nspllc.com/New%20Pages/Business%20Benefits%20of%20Modular%20Software%20in%20Carrier%20Ethernet%20Routers.pdf.

[10] P. Pan et al., "Fast Reroute Extensions to RSVP-TE for LSP Tunnels," RFC 4090.

Carrier-Class Ethernet Service Delivery for Business Access

Using Resilient Packet Rings, Pseudowires, and Multiprotocol Label Switching to Maximize Bandwidth Efficiencies and Minimize Network Costs

Joseph V. Mocerino

Principal Product Marketing Manager
Fujitsu Network Communications

Introduction

Since 1990, more than 1 million synchronous optical network (SONET) elements valued at more than $43 billion have been deployed in North America. These network elements have created the transport backbone for service providers, offering 99.999 percent reliability, guaranteed bandwidth, comprehensive element management, and extremely low latency. SONET achieves this high reliability factor through dedicated reservation of half the available network bandwidth deployed in a diverse path optical ring topology. In the event of a fiber cut or equipment failure, the service will switch over to the alternate protected path in less than 50 ms. High reliability and low latency is the baseline for the voice network. While the SONET infrastructure achieves the performance requirements for voice services, it is inefficient for bursty data networks. The forecast for data services is growing by leaps and bounds. SONET inefficiencies are exacerbated as service demand becomes more packet-based. Data services are deployed using inexpensive Ethernet interfaces of 10/100 Mbps or gigabit Ethernet instead of more expensive and specialized digital signal (DS) 1, DS3, and optical carrier (OC)–n interfaces. Finer granularity than what is available from SONET is needed, and as demand grows, the services need to easily scale. This paper will describe how service providers can capitalize on their investment in SONET equipment while migrating to an efficient Internet protocol (IP) network offering differentiated services.

Next-Generation SONET

To address granularity and scaling issues, the use of next-generation SONET (NG–SONET) techniques are employed. Ethernet over SONET (EoS) capabilities use generic framing procedure (GFP), virtual concatenation (VCAT) and link capacity adjustment scheme (LCAS) to map Ethernet circuits into SONET payloads. GFP ensures compatibility of Ethernet packets mapped into standard SONET synchronous transport signal (STS)–1, STS–3c and STS–12c containers, allowing transport through intermediate, multivendor SONET rings while minimizing the network element upgrade to the outer rings. However, as illustrated in *Figure 1*, mapping Ethernet services into standard SONET containers is not an efficient, straightforward process.

In the 10Base–T example in *Figure 1*, the 10 Mbps Ethernet frames easily fit into the 51 Mbps STS–1 container, but they strand 80 percent of the container bandwidth. The 100 Mbps 100 Base-T circuit is either too big for an STS–1, resulting in dropped packets once the traffic goes above 51 Mbps, or strands at least 33 percent of the 155 Mbps STS–3c container bandwidth. GE circuits, at 1,000 Mbps, are too big for an STS–1 (51 Mbps), STS–3c (155 Mbps) or STS–12c (622 Mbps), resulting in dropped packets once the traffic exceeds the SONET container size. If an STS–48c (2.488 Gbps) is used, more than 50 percent of the bandwidth is stranded. VCAT was created to resolve the misalignment between the Ethernet and SONET rates. VCAT combines a number of VT1.5 (low-order VCAT), STS–1 or STS–3c (high-order VCAT) to create a facility that is a multiple of 1.5 Mbps, 51 Mbps, or 155 Mbps, respectively. The nomenclature used is VT1.5-nv for low order and STS–1-nv, STS–3c-nv, etc., for high-order VCAT. A 10Base–T Ethernet service would use VT1.5–7v, using 10.5 Mbps of container capacity for a 10 Mbps service. A gigabit Ethernet at full rate is STS–3c-8v, or at a fractional rate of 600 Mbps is STS–3c–4v.

VCAT improves Ethernet transport efficiency through more granular scaling than standard SONET containers and concatenation. LCAS dynamically adjusts the "VCAT pipes" when service is upgraded, eliminating the weeks or months a service provider needs to re-provision circuits. If service is disrupted through fiber loss or equipment failure, LCAS will

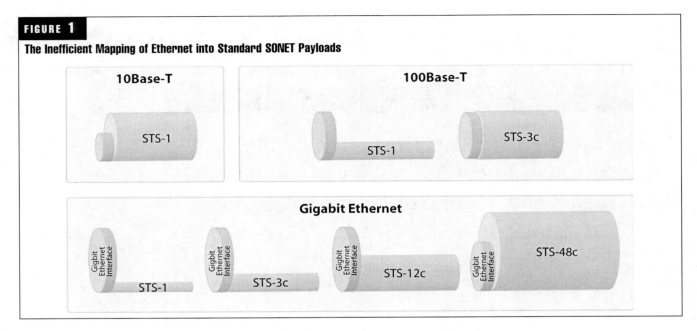

FIGURE 1

The Inefficient Mapping of Ethernet into Standard SONET Payloads

maintain the service without interruption to adjacent units. VCAT and LCAS allow the service provider to offer tiered service pricing based on the granular bandwidth offering.

Ethernet Services

Ethernet Line (E-Line) point-to-point and Ethernet local-area network (E–LAN) multipoint-to-multipoint services are in demand by enterprises in the financial, retail, medical, entertainment, and other industries because of their low cost, ease of use, scalability and manageability. Ethernet services are in greatest demand for a number of applications, including the following:

- Internet access
- Ethernet virtual line and LAN service
- Voice over IP (VoIP)
- Point-to-point Ethernet service

We will now review how service providers can build Ethernet networks over their existing SONET infrastructure based on Metro Ethernet Forum (MEF) connectivity configurations.

Figure 2 illustrates Ethernet private line (EPL) service on a dedicated SONET ring. The customer has three Ethernet virtual circuits (EVCs). Each circuit is used for storage-area networking (SAN), local-area network (LAN) extension and Internet service provider (ISP) access. All services are using guaranteed bandwidth in a point-to-point connection. EPL service is similar to dedicated TDM service using DS1, DS3 or OC-n interfaces but uses lower-cost Ethernet pipes instead. NG-SONET platforms provision EPL services using EoS, which implements high-order (STS–1) and low-order (VT1.5) VCAT to minimize stranded bandwidth and maximize the number of provisioned services on the ring. This design offers more granular bandwidth usage for fractional

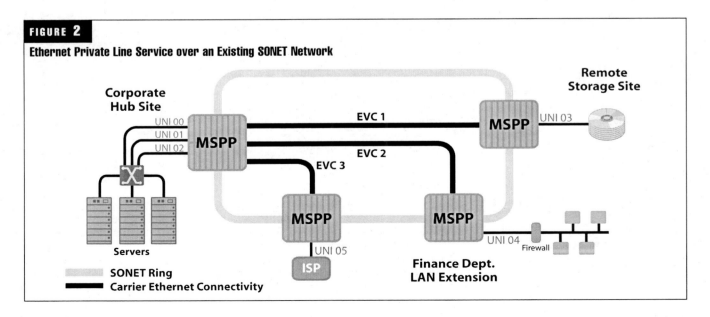

FIGURE 2

Ethernet Private Line Service over an Existing SONET Network

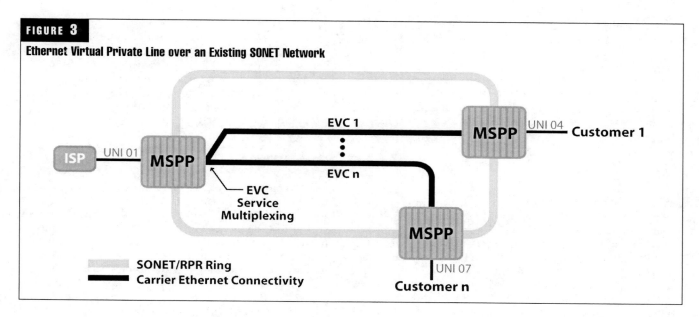

FIGURE 3

Ethernet Virtual Private Line over an Existing SONET Network

Ethernet rates to wire-speed GE. Traffic separation is implemented through the use of separate time slots in the VCAT for each EVC. LCAS is a protocol that dynamically adjusts the size of "VCAT pipes" to meet changing bandwidth requirements for in-service upgrades or downgrades over the existing Ethernet user network interface (UNI). Provisioning and upgrading this service using VCAT and LCAS enable service providers to reduce provisioning time, save a trunk roll and eliminate additional capital expenses (CAPEX) by remotely provisioning service and reusing existing equipment.

The Ethernet virtual private line (EVPL) application in *Figure 3* is similar to the EPL service shown in *Figure 2* except there are multiple customers that are service-multiplexed on a common trunk point. EVPL requires Layer-2

(L2) service functionality with quality of service (QoS) to provide the service multiplexing of EVCs and guaranteed bandwidth or differentiated services for each client. EVPL is similar to frame relay using permanent virtual circuits (PVCs). A classic approach to providing service multiplexing is to use EoS connections with external routers, switches, or bridging elements. Adding an external L2 element to the existing SONET network reduces performance, adds jitter and latency, introduces a single point of failure, and requires a separate element management system (EMS). The use of external L2 equipment is further exacerbated when an Ethernet virtual private LAN (EVPLAN) multipoint-to-multipoint network is provisioned as illustrated in *Figure 4*. Alternatively, the win-win approach is to maintain the SONET attributes of resiliency; performance monitoring; and Operations, Administration, Maintenance and

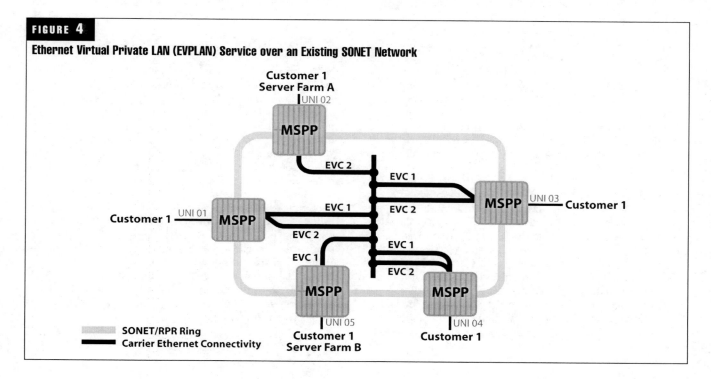

FIGURE 4

Ethernet Virtual Private LAN (EVPLAN) Service over an Existing SONET Network

Provisioning (OAM&P) but process payloads in a L2 packet-based format instead of time division multiplexing (TDM). Multiservice provisioning platforms (MSPPs) support the convergence of NG-SONET equipment, and some offer integrated L2 functionality with QoS options using IEEE 802.17 resilient packet ring (RPR) technology. Using integrated RPR, the MSPP leverages the existing SONET investment while reducing CAPEX and OPEX by eliminating multiple costly core/edge router ports at the central office (CO) through service multiplexing.

SONET migration to RPR

While NG-SONET with EoS works well for EPL applications, its use in EVPL as shown in *Figure 3* and EVPLAN applications as shown in *Figure 4* require multipoint connectivity and service multiplexing. An MSPP with IEEE 802.17 RPR L2 functionality over SONET provides the service provider a smooth migration path from NG-SONET to a converged TDM and packet ring network. A SONET network element with an integrated Ethernet switch using spanning tree protocol (STP) or rapid spanning tree protocol (RSTP) can offer an alternative to RPR in an MSPP. STP or RSTP can provide the connectivity for EVPL and EVPLAN services, but at a cost of high restoration time in the seconds to tens of seconds in the event of a network failure. RPR provides the desired connectivity for carrier-class Ethernet service while maintaining the sub-50 millisecond recovery time. RPR combines SONET performance and survivability with efficient processing of packet traffic. RPR addresses the inefficiencies of SONET by implementing bandwidth sharing, spatial reuse, and statistical multiplexing.

The result of using RPR over SONET instead of EoS is the consumption of significantly less bandwidth and fiber for a given application, and the elimination of external Ethernet switches for EVC service multiplexing. Upgrading to RPR typically requires new service cards and network element software. The network ring bandwidth is assigned customized partitions between TDM and packet bandwidth for RPR. This allows the service provider to maintain existing SONET customers while offering EVPL and LAN services. Ethernet performance monitoring (PM) and alarm management are integrated into the embedded management plane, allowing service providers service level agreement (SLA) management without an additional infrastructure. The result is a converged packet and TDM platform offering the reliability, availability and OAM&P of SONET with the robust data efficiencies and multipoint connectivity of packet-based services.

RPR uses a dual counter-rotating ringlet topology and sub-50-millisecond steering protection to maintain services upon a fiber cut or line card failure. RPR QoS provides differentiated Ethernet services for guaranteed, partially guaranteed and best-effort services. Class A, high-priority QoS supports strictly bounded delay and jitter tolerances with user-provisioned, non-reclaimable, committed information rate (CIR-1) bandwidth settings. Class A is used for provisioning voice and real-time video applications. When CIR-1 is used, bandwidth is reserved and guaranteed on both RPR ringlets. Class B supports applications less sensitive to delay and jitter such as non-real-time video and virtual private network (VPN) services. Statistically, reclaimable CIR (CIR-2) and excess information rate (EIR) options are supported on Class B traffic. The CIR-2 option reserves and guarantees bandwidth on both RPR ringlets when in use, otherwise the bandwidth can be reused by EIR traffic in the RPR. Class C is for lower-priority service classes supporting best-effort applications that do not need bandwidth guarantees, such as basic Internet access.

When data traffic is partially used or not present from one Ethernet virtual circuit (EVC), another EVC in the same RPR will use the bandwidth where the sum of the path bandwidth is partially or completely shared using QoS prioritization and policing. In contrast, SONET uses TDM where EVC bandwidth is always dedicated and the sum of the provisioned EVC bandwidth cannot exceed the sum of the path bandwidth. RPR offers greater utilization of path bandwidth and optimizes efficiency for bursty data applications to allow more Ethernet services to be provisioned on the OC-12, OC-48 or OC-192 ring than SONET.

RPR service classes, QoS, and traffic shaping allow service providers to offer differentiated Ethernet services, which include tiered pricing and provisioning options. RPR QoS bandwidth options can provide a 70 percent or greater increase in the number of bursty Ethernet services on a given shared RPR network as compared to a fixed TDM approach. RPR packet connectivity eliminates the need for external L2 Ethernet switches or bridges when deploying EPL, EVPL and E-LAN multipoint-to-multipoint services. Integrated Ethernet L2 switching and bridging reduces capital expenses and improves service performance through less jitter and latency for mission critical applications such as VoIP and non-real-time video. RPR topology provides a more than 40 percent reduction in network line interfaces than a typically sized carrier-class (fully protected) hub-and-spoke or mesh network topology. Reduced equipment requirements result in lower CAPEX and fewer single points of failure, which also reduces OPEX.

MAN–to–MAN Connectivity Using MPLS

RPR connectivity within the metropolitan area network (MAN) provides multipoint connectivity, resiliency, and differentiated services for E-Line and E-LAN services. For service connectivity between RPRs, several techniques are used to migrate the installed base to a full network connectivity model. Service performance is bounded by IEEE RPR 802.17, which defines a single ring with a maximum circumference of 2000 km. The simplest interconnection of MAN rings would be at the optical hub using L2 bridging with Q-in-Q tagging to maintain traffic separation. For service connectivity between optical hubs, the RPR VLANs will be terminated and interconnected to the multiprotocol label switching path. An MPLS path is established for a given sequence of packets via its control plane as illustrated in *Figure 5*. Trunk points are established on the optical hub using GE circuits. Protection switching is implemented using 802.1ad aggregation. A pair of GE ports configured for protection switching will be needed for each RPR to interconnect optical hubs.

FIGURE 5

Resilient Packet Ring Multiservice Provisioning Platform Migration to Multiprotocol Label Switching

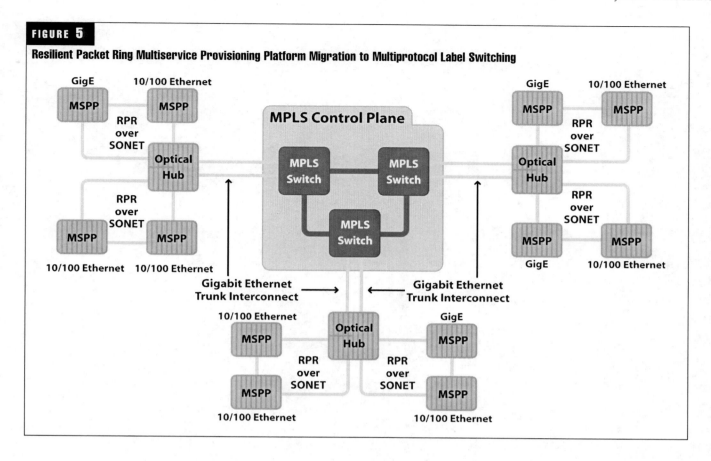

Legacy business traffic such as frame relay and asynchronous transfer mode (ATM) is carried across the MPLS network using pseudowire emulation edge-to-edge (PWE3) encapsulation.

With this unified MAN–to–MAN IP infrastructure, service providers can continue to realize benefits from legacy services while addressing new high-speed packet services. The RPR/MPLS network can provide bandwidth relief with differentiated QoS for multi-ring networks, which maintains most of the installed investment of SONET equipment while providing cost-effective carrier-class Ethernet services to business customers.

References

IEEE Draft P802.17/D3.3, "Part 17: Resilient Packet Ring (RPR) access method & physical layer specifications," 21 April 2004.

Metro Ethernet Forum Technical Specification MEF 6, "Ethernet Service Definitions- Phase 1," June 2004.

Metro Ethernet Forum Technical Specification MEF 9, "Abstract Test Suite for Ethernet Services at the UNI," Oct 2004.

VPLS.ORG: VPLS Standards (vpls.org/vpls_standards.pdf), 2004.

Cost-Effective and Practical Implementation of the Gigabit Ethernet for Home and Small-Business Applications

Subbarao V. Wunnava
Professor, Electrical and Computer Engineering
Florida International University

Hilda Palencia
Research Assistant, Electrical and Computer Engineering
Florida International University

Jaime Montenegro
Research Associate, Electrical and Computer Engineering
Florida International University

Abstract

Electronic commerce, multimedia entertainment, on-line education, and similar applications have pushed the need for adapting high-speed gigabit Ethernet (GE), even for homes and small businesses. There is considerable confusion and concern about updating the conventional 10 or 100 Megabit Ethernet to GE, especially in terms of cost, performance, and compatibility with the existing networking modules. There is abundance of literature on GE products and services, which often leads to more confusion in the minds of the home and small-business owners rather than helps them. Also, there are no clear guidelines as how to assess the improvement in terms of reliability, security, performance, and cost-effectiveness, when GE is adapted, in place of the conventional Ethernet [1, 2, 3, 4].

In this article, the authors will present the important issues of migration into the GE environment and discuss the cost-performance issues. Also, the authors will provide some practical implementation methodologies based upon the simulation results of GE. Some invisible factors that affect the performance of GE are latency or delay and point-to-point and multipoint traffic flow. These issues are currently being studied at the VLSI and Networking Laboratories at Florida International University. The authors will also provide the results of these investigations. The authors would like to thank Mr. Magno Guillen for helping them in the OPNET simulations.

Introduction

Due to significant technological advancements in computing power and storage systems, it is now easier to support new and existing network and computer applications with high-bandwidth data, high-resolution graphics, and other complex and rich multimedia data. One of the most invasive networking technologies that currently comply with this demand is GE. It is the latest version of Ethernet. It offers 1,000 Mbps (1 Gbps) of raw bandwidth, which is 100 times faster than the original Ethernet, yet is compatible with existing Ethernets, as it uses the same carrier sense multiple access with collision detection (CSMA/CD) and media access control (MAC) protocols [5]. The original Ethernet was developed as an experimental coaxial cable network in the 1970s by Xerox Corporation to operate with a data rate of 3 Mbps using CSMA/CD protocol for local-area networks (LANs) with irregular but occasionally heavy traffic requirements. Success with that project attracted early attention and led to the 1980 joint development of the 10 Mbps Ethernet Version 1.0 specification by a three-company consortium: Digital Equipment Corporation, Intel Corporation, and Xerox Corporation [5]. In March 1996, the IEEE 802.3 committee approved the 802.3z Gigabit Ethernet Standardization project. At that time, as many as 54 companies expressed their intent to participate in the standardization project. The Gigabit Ethernet Alliance was formed in May 1996 by 3Com Corp., Bay Networks Inc., Cisco Systems Inc., Compaq Computer Corp., Granite Systems Inc., Intel

Corporation, LSI Logic, Packet Engines Inc., Sun Microsystems Computer Company, UB Networks, and VLSI Technology. Physical layer standards include 1000BASE-T, 1 Gbps over Cat-5e copper cabling, and 1000BASE–SX for short to medium distances over optic fiber. The various layers of GE protocol architecture are shown in *Figure 1*. The gigabit media internet interface (GMII) is the interface between the MAC layer and the physical layer. It allows any physical layer to be used with the MAC layer. It is an extension of the media independent interface (MII) used in fast Ethernet. It uses the same management interface as MII. It supports 10, 100, and 1,000 Mbps data rates. It provides separate 8-bit wide receive and transmit data paths, so it can support both full-duplex as well as half-duplex operation [5, 6, 7].

GE Considerations

This article discusses the design and development of a GE LAN. Two such LANs are interfaced through bridges. It includes an identification of the required modules and relative network cost analysis. The network will be simulated using OPNET Modeler (which is a discrete event simulator) for performance analysis. The GE LAN environment that was designed is shown in *Figure 2*.

OPNET Implementation of GE
Figure 3 shows the overall OPNET implementation of GE LAN. This illustrates the connectivity of two LANs with the appropriate bridges, firewall, and Internet connection. LAN1 and LAN2 are linked using two *ethernet2_bridge_base* models. The gateway used for connection between the Internet cloud (*node_1*) and the LANs is a firewall router

(*node_0: ethernet8 –slip2-firewall*). A more detailed description for each element is as follows:

- *Ethernet2_bridge_base*: They support up to two Ethernet interfaces. Protocols used are spanning tree bridge protocol (Institute of Electrical and Electronics Engineers [IEEE] 802.1D), Ethernet, fast Ethernet, and GE. It interconnects with two Ethernet connections at 10, 100, or 1,000 Mbps [8].

- *Node_0: ethernet8 –slip2-firewall*: It represents an Internet protocol (IP)–based gateway with firewall features and server support. IP packets arriving on any interface are routed based on their destination IP address. Protocols used are transmission control protocol (TCP), routing information protocol (RIP), user datagram protocol (UDP), IP, Ethernet, fast Ethernet, GE, open shortest path first (OSPF), border gateway protocol (BGP), and interior gateway routing protocol (IGRP) [8].

- *Internet cloud (node_1): ip32_cloud*: It represents an IP cloud supporting up to 32 serial line interfaces at a selectable data rate through which IP traffic can be modeled. IP packets arriving on any cloud interface are routed based on their destination IP address [8].

Different objects or nodes create the inside of the LAN1 and LAN2 subsets. These components are put together as illustrated in *Figure 4*. This scenario was set up with the following objects:

- *LAN models*: A model of a 64-station LAN was obtained using a switch that connects 64 stations. Two

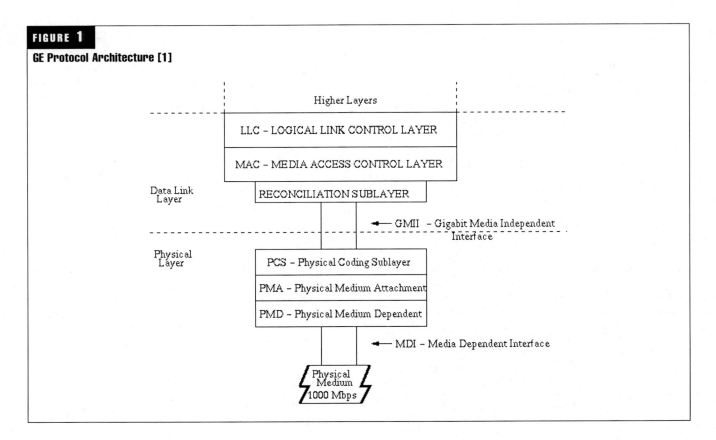

FIGURE 1

GE Protocol Architecture [1]

FIGURE 2

Overall Network Diagram

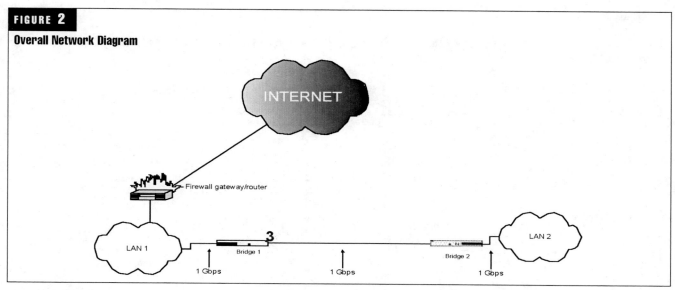

FIGURE 3

GE LAN

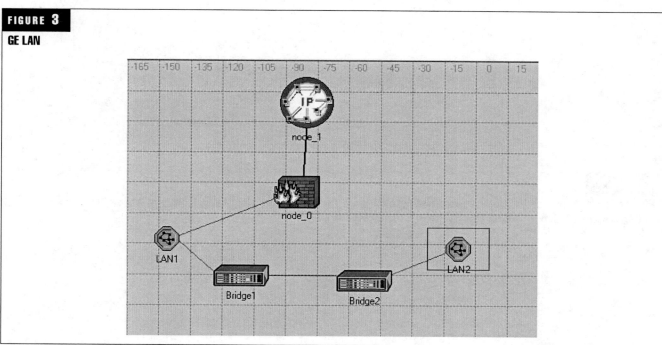

of these stations are used as file servers (FS1 and FS2); two nodes are connected to represent the print servers (PS1 and PS2). To represent a GE LAN, a 1000BaseX_LAN object topology was used.

- *Profile configuration*: It is applied to a workstation, server, or LAN. It specifies the applications used by a particular group of users. The profile configuration node can be used to create user profiles. These user profiles can then be specified on different nodes.

- *Application configuration*: It may be any of the common applications (e-mail, file transfer) or a custom application defined. Eight common ("standard") applications are already defined: database access, e-mail, file transfer, file print, telnet session, videoconferencing, voice over IP (VoIP) call, and Web browsing. *Figures 4* and *5*

depict the application configuration menu options and the used values for LAN1 and LAN2.

- *Links*: Gig_EthCh_adv duplex link represents an EtherChannel connection operating at n * 1,000 Mbps (or 1 Gbps) in full-duplex mode.

Figure 6 shows the OPNET scenario for the interconnectivity of the LANs with two print servers and two file servers. For the model analysis, we used 64 stations (including the file servers). Such an environment is typical in small-business and home applications [6].

Simulation Results of GE

The implemented network was simulated and several parameters were obtained. *Figure 7* shows the global or the

FIGURE 4

OPNET Applications Used for Traffic Generation in LAN1. Fifteen Applications Were Simulated. Main Load Was an FTP Operation Using 100 MB Files.

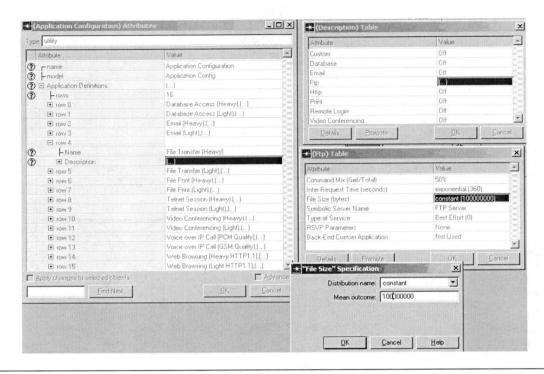

FIGURE 5

OPNET Applications Used for Traffic Generation in LAN2. Fifteen Applications Were Simulated. Main Load Was an FTP Operation Using Files That Follow a Uniform Distribution of 5 to 100 MB File Size.

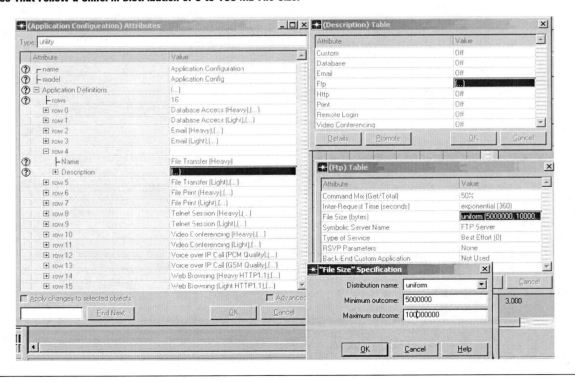

FIGURE 6

OPNET Scenario for LAN1 and LAN2—64 Stations (Including Two File Servers), Two Print Servers, Profile Configuration, and Application Configuration

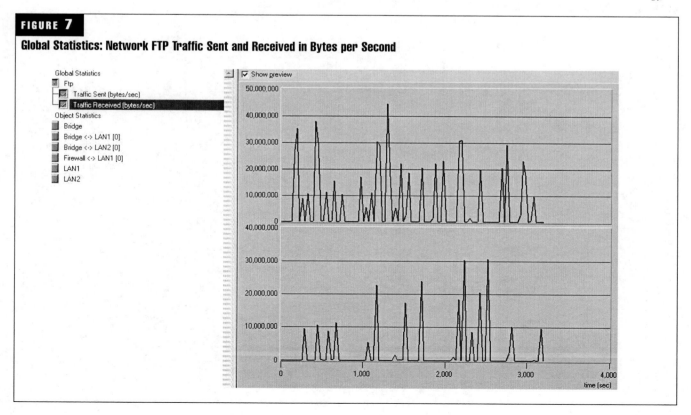

overall network results for sent and received FTP traffic. The chart shows oscillated traffic of sent and received files, having max peaks of 45 Mbytes/sec (360 Mbps).

Different parameters were monitored for the simulated model. *Figure 8* shows the network delay and throughput for LAN1. The delay shows a very variable behavior from 5 to 70 ms. The inbound and throughput traffic show a maximum peak of almost 41,000,000 bits per second.

Figure 9 shows network delay, inbound and through traffics in bits per second for LAN2. In this case there is a maximum delay of 95 ms. The maximum traffic observed was of 29,000,000 bits per second.

Figures 10 and *11* illustrate the point-to-point throughput for the file servers FS1 and FS2 in LAN1 and LAN2, respectively. Note that the difference in the shape of the traffic pattern among the LAN2 and LAN1 is due to the statistical model used for the traffic generation in each LAN.

Finally, *Figures 12* and *13* show the traffic throughput in the bridge connection and between the LAN1 and the firewall.

Network Relative Cost Analysis

While the software cost depends upon the capabilities needed, each gigabit LAN will have some minimal hardware elements and associated costs. Each LAN technology is

FIGURE 7

Global Statistics: Network FTP Traffic Sent and Received in Bytes per Second

FIGURE 8

LAN Statistics: Delay and Traffic for LAN1

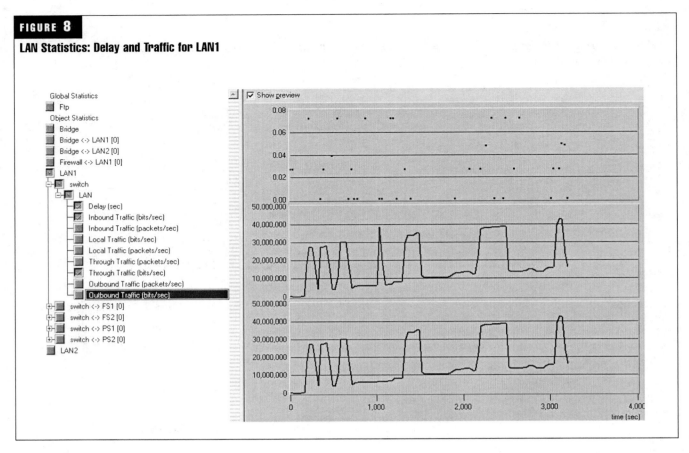

FIGURE 9

LAN Statistics: Delay and Traffic for LAN2

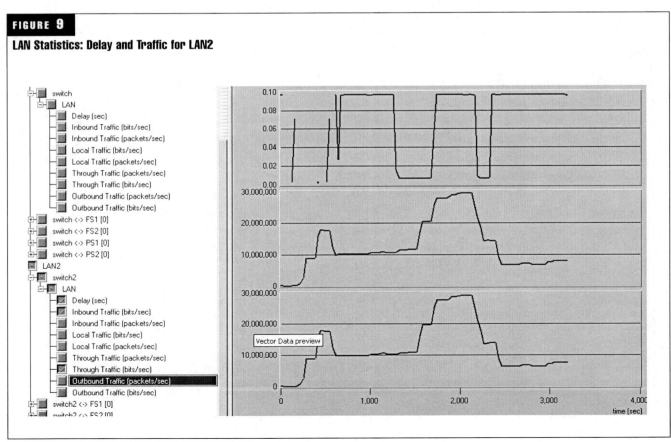

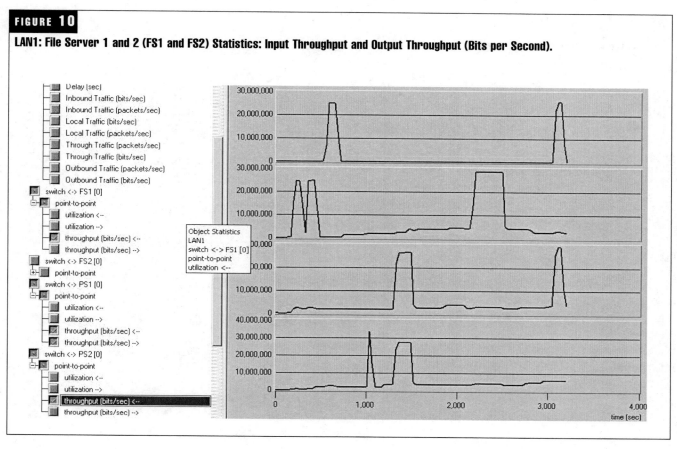

FIGURE 10

LAN1: File Server 1 and 2 (FS1 and FS2) Statistics: Input Throughput and Output Throughput (Bits per Second).

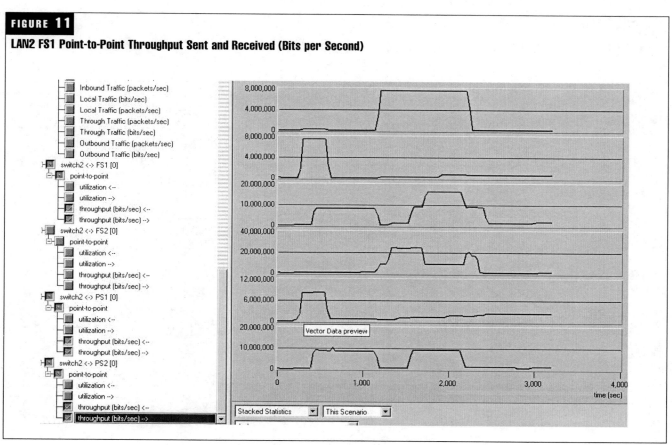

FIGURE 11

LAN2 FS1 Point-to-Point Throughput Sent and Received (Bits per Second)

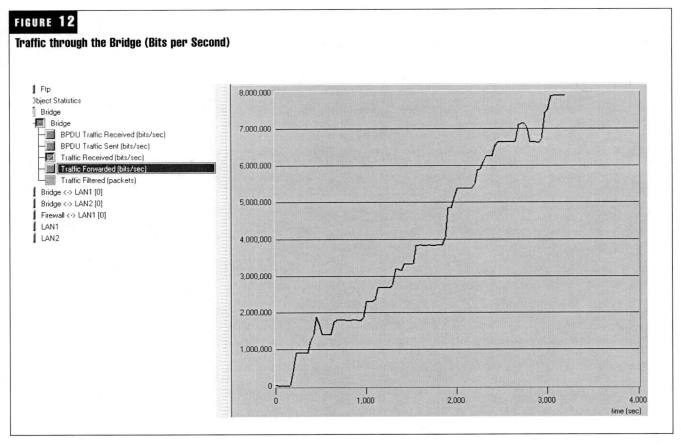

FIGURE 12

Traffic through the Bridge (Bits per Second)

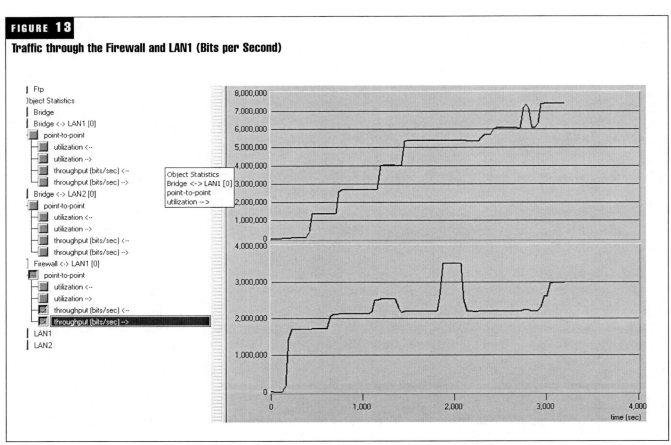

FIGURE 13

Traffic through the Firewall and LAN1 (Bits per Second)

designed for a specific combination of speed, distance, and cost. When designing a network, engineers choose a combination of capacity, maximum delay, and distance that can be achieved at a given cost. To help save expense, LAN technologies usually use a shared communication median such as a shared bus or ring. The developed network has detailed engineering requirements and capacities of the equipment to ensure accurate cost modeling of their equipment.

Required Hardware Modules

- *Four catalyst switches*: Two switches of 48 ports or one switch of 120 could be used fit 62 workstations, two file servers, and two print servers.

- *Two bridges*: One bridge for each LAN. Fiber converters are suggested for this type of network.

- *Two print servers and two file servers*: A total of four servers are needed to have a storage capacity of 1 Terabyte.

- *One router for firewall*.

There are several companies, including Cisco, 3Comm, Trendnet, NETGEAR, and HP, that provide the above-mentioned hardware units. The prices vary, especially depending upon the number of units purchased.

Relative Summary of Costs

With system on chip (SoC) technology becoming very cost-effective, the price of the gigabit LAN hardware is continually going down. The actual cost numbers should be obtained from the manufacturers' Web sites and the representatives. However, the authors have found that the cost-effective solution lies in having a system with standard interfaces. In that process, the authors suggest the following elements (choose one from each category):

Switches
1. Cisco Catalyst 3560G-48TS SMI – switch – 48 ports
2. HP ProCurve Routing Switch 9308m – 120 ports
3. 3com Switch 8807 – 120 gigabit ports

Bridge
1. Trendnet TFC-1000MSC multimode fiber converter bridge

Servers
1. HP ProLiant ML370
2. Dell PowerEdge SC1420

Routers
1. NETGEAR ProSafe 48 router
2. HP/Compaq – ProCurve 2650-PWR – 48-port switch/router

Workstations
1. Dell or HP, or Compaq, HP, or Apple Systems with 1 gigabyte RAM and 140 gigabyte hard drives, with standard GE interface

Conclusion

A GE LAN was designed, simulated, and implemented using OPNET Modeler. Various parameters were analyzed, which showed, on most occasions, normal behaviors of utilization and transfers, following the CSMA/CD protocol expectations. The simulation results also showed how the printer servers are being used more than the file servers, especially in a small-business environment. In addition, a relative cost analysis of an actual LAN was discussed, taking into consideration several configurations from different manufacturers. The final cost has to be acquired according to the system specifications and economical factors. Where there is a need for redundancy and fault tolerance, it is appropriate to use switches with 48 ports. However, the authors also observe that using multiple 48-port switches in place of a 120-port switch at times slows down the system. This may be attributed to the actual routing and buffering mechanisms.

Bibliography

1. Yossi Saad, Vice President, Product Marketing, Actelis Networks, Inc., "Ethernet over Copper Generates New Revenue Streams to Service Providers," Annual Review of Communications, IEC Publications, 2004, www.iec.org/pubs/print/aroc57.html.
2. Interoperability Laboratory, "Gigabit Ethernet Consortium Special Report," June 29, 2005, www.iol.unh.edu.
3. Special report from Intel Corporation "Intel Network Connectivity Gigabit Solutions," Networking Division of Intel, www.intel.com/network/connectivity/solutions/gigabit.htm, September 2005.
4. Special report by Cisco, "Deploying the Gigabit Ethernet to the Desktop," October 2005, www.cisco.com/en/US/tech/tk389/tk214/technologies_white_paper09186a00801a7595.shtml.
5. Moorthy, V., "Gigabit Ethernet," 2005, www.cse.wustl.edu/~jain/cis788-97/ftp/gigabit_ethernet/index.htm#BIB.
6. Hyperlink Technologies, www.hyperlinktech.com/web/hg2408p.php.
7. Alexander Lackpour, Mohsen Kavehrad, Scott Thompson, Architecture and Predicted Performance of an IEEE 802.11b-like Wireless Network, Center for Information and Communication Technology Research, PENState University, 2001.
8. OPNET Tutorial Manual. OPNET Technologies, 2001.
9. OPNET lab manual By William Stalling, ISBN 0-13-148252-1.
10. Cisco Catalysts: www.cisco.com/now/ipcommunications/?sid=118527_979.
11. Hewlett-Packard: welcome.hp.com/country/us/en/prodserv.html.
12. 3com: www.3com.com/index2.html.
13. Trendnet: www.trendware.com.
14. Dell: www.dell.com.

GE ADM–Based Ethernet Service Transmission Solution

Michael Xiong

Marketing Manager, Optical Network Division
Huawei Technologies Co., Ltd.

Today, both services and networks are more and more Internet protocol (IP)–based. Considering such issues as the development trend of metropolitan-area networks (MAN), service capacity, transmission distance, and performance-to-price ratio, gigabit Ethernet (GE) has become the mainstream interface for metropolitan broadband data network interworking. In face of the mass GE interworking, it becomes more and more important to create a solution to transmission and build a manageable, operational, low-cost, and quality of service (QoS) guaranteed carrier-class metropolitan network.

Introduction

With the construction of asymmetric digital subscriber line (ADSL), fiber-to-the-home (FTTH), and IPTV networks, operators are able to provide abundant service experiences, including Internet browsing, movie downloading, voice over IP (VoIP), and network gaming for public users, and videoconferencing, remote education, and virtual private networks (VPNs) for business users.

With the large expansion of backbone and access network capacities, operators are shifting their competition focus from backbone networks to metro networks. Considering development trends, service capacity, transmission capacity, and performance-to-price ratio of metro networks, GE is the mainstream interface for metro broadband data service interworking. How to effectively solve the service transmission problems and create a manageable, operable, and low-cost carrier-class metro network with a QoS guarantee is a growing concern for operators.

Transport Problems of Metro Ethernet Services

There are many solutions for metro Ethernet service transmissions, but they fail to reach a high performance-to-price ratio. Metro Ethernet service transmission imposes the following challenges:

- Low use of fiber source, small network capacity, limited transmission distance, and expensive long-haul transmission
- Complex network structure, limited virtual local-area network (VLAN) resources and media access control (MAC) address resources, and limitation of network upgrade and expansion

- End-to-end QoS guarantee problem
- Carrier-class protection switching problem
- Protocol independence and transparent transmission
- Fast support for unsure services
- Secure isolation between users
- Network performance monitoring and internal operations, administration, and maintenance (OA&M) capabilities

GE ADM Solutions

GE ADM versus Metro WDM

Now, the simple way of improving line rate of transmission alone is to no longer meet the bandwidth requests of services. By using wavelength division multiplexing (WDM) technology in metropolitans, operators can improve their fiber capacity and solve the problem of transmitting signals of multiple wavelengths over one fiber and add/drop services directly in the optical layer. However, it fails to handle the problem of single-wavelength efficiency, which can be solved by GE add/drop multiplexer (ADM) technology. With GE ADM, operators can add/drop GE services freely, and GE of multiple nodes can share a single wavelength. The perfect combination of GE ADM and metro WDM offers a new solution for IP digital subscriber line access multiplexers (DSLAMs), IPTV, and GE VPNs.

According to the different kind of networks, metro WDM system can provide the proper transmission solution from the access or convergence layer to the core layer. All these instruments share the major functional modules, software or hardware platforms, and reduce capital expenses (CAPEX) and operational expenses (OPEX).

GE ADM–Based Metro WDM Networking

The difference between a GE ADM system based on metro WDM and a traditional WDM system lies in the GE ADM system using the architecture of optical ADM (OADM)/reconfigurable OADM (ROADM) and GE ADM. The use of wavelength conversion units and multichannel conversion units can access services. OADM/ROADM technology can add/drop optical-layer services and make these services pass through. A GE ADM matrix can add/drop and groom GE sub-wavelengths. The combination of an optical-layer channel-level service grooming function and an electrical-layer GE–channel–level service grooming function can implement a fiber-sharing function and a wavelength-sharing function. *Figure 1* illustrates the system hierarchy of the metro WDM system.

OADM/ROADM

Metro WDM provides OADMs with fixed wavelengths, including parallel mode, serial mode, and their combination. These OADM modules can be smoothly upgraded and expanded by overlapping or cascading. To simultaneously improve flexibility of wavelength configuration and service adding/dropping, it adopts the ROADM and tunable lasers to enhance the metro WDM network flexibility and reduce the cost of standby devices.

GE ADM

The improvement of wavelength utilization can effectively reduce network costs. The unique GE ADM technology defines GE services as the basic granularity in the wavelength for adding/dropping GE services directly. GE ADM technology multiplexes multi–GE services into a wavelength. Through the embedded cross-connect grooming unit, it can add/drop GE services flexibly or let GE services pass through. In that case, it achieves transparent grooming of GE granularity, and provides flexible and reliable configuration plan and strong service convergence as well as grooming capabilities in the metro network.

GE granularity in GE ADM is quite similar to VC12 in the synchronous digital hierarchy (SDH), and its wavelength is similar to VC4 in the SDH. The total capacity of the WDM system is similar to the line rate of the SDH.

Advantages of GE ADM

Through the grooming unit of GE ADM, GE of different nodes might share one wavelength in which 75 percent of wavelengths are reduced. This has several advantages, including the following:

- It can reduce the cost of OADM.
- It provides excellent GE channel protection and GE channel shared protection.

- It provides end-to-end GE service configuration and fast service provisioning.
- It provides flexible upgrade and expansion mode and GE service grooming.

Flexible Service Grooming

The metro WDM system uses OADM and GE ADM technology to achieve flexible upgrade, expansion, and service grooming for GE channels. When using fixed OADMs to create wavelength channels, various numbers of GE service channels might be configured at different nodes according to network service requirements, activate/deactivate or add/delete GE services. With the complete remote software drive, no other service channels are affected.

By using ROADM and tunable lasers, the metro WDM system can improve the flexibility of wavelength assignment. However, it fails to dynamically groom or assign GE services. By using ROADM, tunable lasers, and GE ADM technology, the metro WDM system might flexibly provide or groom GE sub-wavelengths and wavelength-level services and provide expressways for operators of data services.

Applications of GE ADM

GE Transmission in ADSL/FTTH

The GE ADM technology offers the complete DSLAM transmission solution from the access layer to the core layer. With the ability to effectively transport Ethernet services from the edge nodes to the center node, it can be applied either to single-homing or dual-homing data service networks.

GE ADM technology can provide transparent transmission and cross-connect grooming of GE services. It is able to fulfill the purpose of adding/dropping GE services directly to/from the fiber or wavelength. With the customized multichannel GE service convergence unit, it can multiplex low-speed Ethernet services into the high-speed wavelength to improve wavelength use. GE services at different nodes can share one wavelength, reducing the network cost.

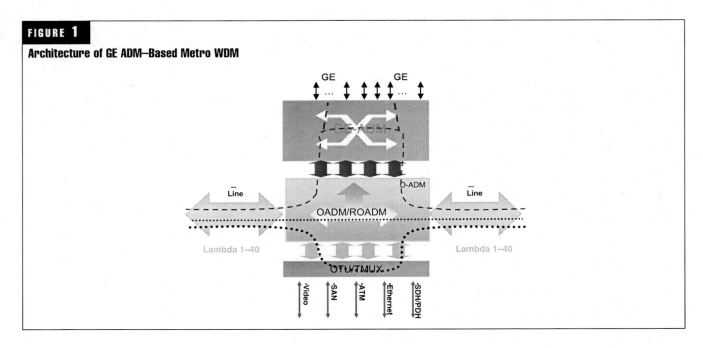

FIGURE 1

Architecture of GE ADM–Based Metro WDM

A DSLAM transmission solution based on GE ADM reduces the number of hops for electrical regeneration or store and forward, using minimum wavelengths and optical amplifiers to bear Ethernet services. In addition, it directly adds/drops wavelengths at the optical layer and adds/drops GE services at the electrical layer. It not only expands the network capacity, but also enhances network reliability and OADM capability.

GE VPN

GE ADM technology can realize multiplexing of multi–GE services of different nodes into one wavelength. It features flexible GE channel configuration, grooming, and routing in the wavelength. It can be used to achieve GE VPN. Users can create multiple VPNs based on GE sub-wavelength. In this case, users can add to or remove from the VPN group, fulfill physical isolation in the WMD system, and realize large capacity and high security.

GE Multicast Applications for IPTV

IPTV includes rich services, which might be grouped into unicast (video on demand [VoD]) and multicast (i.e., broadband TV) by the type of traffic flow. IPTV services need numerous real-time and interactive features. Therefore, the network quality by using multilayer storage forwarding mode is uncontrollable. The GE ADM technology can conveniently achieve GE multicast and broadcast functions.

The metro WDM sends all GE channels to the downstream node, which uses GE ADM to perform transparent cross-connect duplication for GE channels and then transmit the duplicated GE to the next node. In multicast or unicast mode, the metro WDM system provides wavelength-level protection and GE–channel–level protection.

Application Cases

By now, metro WDM equipment has been used all over the world, including the case of the core metropolitan data network of global technical services (GTS) in Budapest, Hungary, and DSLAM transport network of Novis in Portugal.

Novis is a very important fixed-network operator in Portugal. With the sharp increase of broadband data services on the Internet, Novis had difficulties constructing a nationwide broadband data transport network, especially in some larger cities such as Lisbon, the capital, and Porto, the famous harbor city, where there are a lot of symmetric digital subscriber line (ADSL) subscribers, whom Novis offered abundant services such as IPTV, network video, and broadband Web browser. Therefore, a lot of GE channels of broadband access equipment should realize their efficient and reliable internetworking. Considering the service development and the requirements of network expansion and upgrade, there were four GEs in each DSLAM that needed to be uploaded. Obviously, traditional transport networks could not implement this requirement of service provisioning of Novis. After multilateral discussions, Novis chose an OADM and GE ADM hybrid solution to construct its GE ADM–based Ethernet transport platform from the access to the core layer. With this platform, Novis can realize fast GE service provisioning and provide ideal protection in the optical layer to save the optical fibers and reduce the cost of construction, maintenance, and upgrade.

FIGURE 2

Parts of DSLAM Transport Network in Lisbon, Portugal

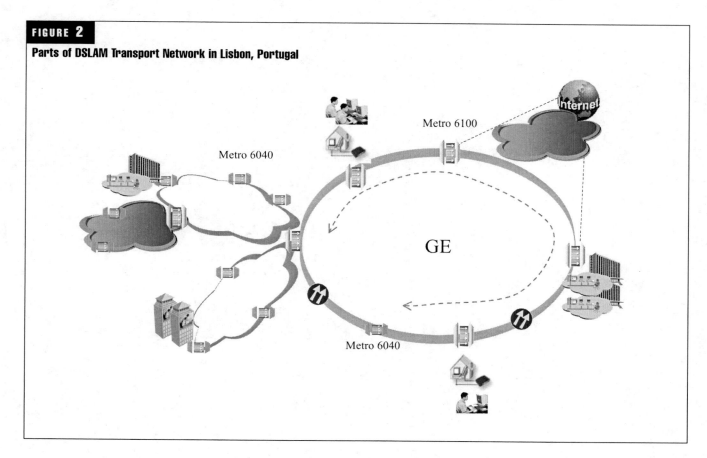

Figure 2 illustrates some parts of the network. In Lisbon, OptiX Metro6100s are used in large-capacity offices in metropolitan-area transport networks, while OptiX Metro6040s are used in small ones. Furthermore, a unified hardware/software platform and network management system have also been adopted, which greatly reduced the construction cost of the whole network.

Conclusion

By using a GE grooming and wavelength grooming function, GE services and wavelengths can be flexibly added to or dropped from the fiber directly, and GE services or wavelength channels can be provided from end to end quickly in the metro network. The combination of optical-layer protection and GE channel-layer protection provides excellent multilayer protection mechanisms, thus ensuring QoS and network survivability capabilities and constructing efficient Ethernet service transmission solutions such as IP DSLAM, IPTV, and GE VPN.

Triple Play

Transport Considerations for Triple-Play Deployments

Paul Forzisi

Vice President of Marketing
White Rock Networks

Introduction

The term "triple play" has come up frequently in relation to communication services over the past couple of years. While there does not seem to be much disagreement that the trio of voice, video, and data services yields more revenue and promotes customer retention, the implications it has on access network architecture depends on the kind of service provider a company traditionally has been. For residential services, the dominance of each service for wireline and cable providers has generally broken down as follows:

Wireline carriers	Cable providers/multiple-system operators (MSOs)
1. Voice telephony	1. Video
2. Internet	2. Internet
3. Video	3. Voice telephony

From the above, we can see why the competitive meeting point between wireline and cable carriers has, to this point, been Internet services. However, while the Internet provided revenue growth opportunities for wireline and cable companies alike, the expansion of cable companies into voice telephony services encroaches on the wireline carrier's core business. A natural way for wireline carriers to defend this key business is to take an offensive approach and consider the delivery of video services in competition to the cable providers/MSOs. The bundling of two or more services has already proven to increase customer retention and reduce churn, especially when service bundle discounts are applied. Triple play extends this approach even further to the one-stop-shop concept.

Apart from the obvious external threats, wireline carriers have seen decreasing voice-line revenues from internal competition as broadband access continues to supplant the narrowband Internet access model. Many customers have disconnected second lines to the home typically dedicated to dial-up Internet service. Similarly, we have seen e-mail replace fax service, creating yet another reason for fewer voice lines in the home. The end result is that digital subscriber line (DSL) access service charges are partially offset by decreasing narrowband service revenues through internal cannibalization.

However, in the march toward bundled service offerings by all service providers, change favors the wireline carriers. The telephony services contested by the cable providers/MSOs are already considered a ubiquitous utility by customers, and they offer carriers limited margins. In addition, hybrid fiber/coax (HFC)–based networks currently fail to provide lifeline services, thereby limiting the addressable market to second lines, a market that is already shrinking. Video, on the other hand, increasingly represents entertainment to the mass market and in turn tends to attract higher fees and margins for the associated services. Internet service is settling in somewhere between the two, with the relative value dependent on each customer's usage model.

Beyond the competitive positioning, video is a way for wireline carriers to offer differentiated services based on customer preferences. The newest technologies for wireline video delivery now support video on demand (VoD) and other interactive services in addition to standard broadcast video.

Broadband-Ready Access Networks: The Right Direction

When it comes to the investment in broadband Internet access infrastructure, the news is good for service providers considering video services delivery. The modern method of delivering packet-based video over wireline networks fits well with the existing broadband access architecture. While video extends the challenges posed to access network planners over just broadband Internet service alone, the general move to broadband can be considered a step in the right direction.

Technologies such as asymmetric DSL (ADSL) in the last mile, the increased packet routing infrastructure placed in central office (CO) and point of presence (POP) locations, and broadband fiber lines placed deeper into the access network partially prepare wireline providers for the move to video. Such a move would allow service providers to more fully leverage the broadband assets placed into the access network with revenues from additional services.

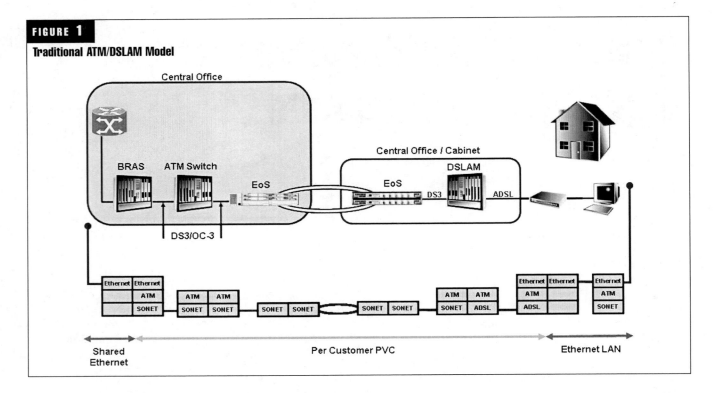

FIGURE 1

Traditional ATM/DSLAM Model

Figure 1 shows the network elements typical of a broadband access network used for the delivery of high-speed Internet services. The traditional model uses asynchronous transfer mode (ATM) permanent virtual circuits (PVCs) from the DSL modem in the customer premises all the way through to the broadband–remote access server (B–RAS) at the POP. The trunk port of the DSL access multiplexer (DSLAM) hands off an ATM user network interface (UNI) over various synchronous optical network (SONET) interfaces. Traffic within the UNI is aggregate traffic from all customers, but each customer is mapped into the UNI as an individual ATM PVC.

This model works fine for small-scale deployments. But as customer numbers grow, the large number of PVCs becomes extremely difficult to manage. ATM cross-connects must be configured in the DSLAM and at all intermediate ATM aggregation nodes before the service reaches the B–RAS. At the B–RAS, the ATM service is terminated along with the user-specific point-to-point protocol (PPP) session. Then the user traffic is passed onto the Internet backbone as IP–over–Ethernet.

The ATM model uses switches to aggregate services from multiple DSLAMs. ATM services are typically carried over some form of SONET interface, thus a large-scale network of this type will use many expensive ATM switches with expensive ATM/SONET interfaces.

Outgrown ATM Networks Turn to Ethernet for Help

Lack of scalability with the ATM model has spurred a recent trend in DSLAMs to provide the aggregate output on an Ethernet trunk port. IP DSLAMs are unlike their ATM cousins; they do not require an individual cross-connect and

PVC for each customer on the trunk-side port. Instead, the IP DSLAM de-encapsulates Ethernet traffic from the ATM stream and hands it off as native Ethernet to the transport network, as shown in *Figure 2*.

Traffic in the local loop between the customer premises and the DSLAM remains in ATM format. This allows the service provider to maintain the quality of service (QoS) necessary for each service when multiple services share a common access line. On the trunk side of the DSLAM, services are generally fed to the DSLAM on individual ports, making the DSLAM the convergence point for Internet and video services. This in turn does not require that a relative QoS mechanism be provided in the transport portion of the access network. The absolute QoS in the access network is provided by the underlying SONET infrastructure.

As mentioned above, the ATM model inherently creates a virtual private connection between the end user and the B–RAS. In turn, the ATM connection carries a PPP session encapsulated in Ethernet that is terminated by the B–RAS. The PPP protocol provides a secure, point-to-point, single-user interface to the B–RAS, where it is then stripped away before forwarding the IP packets into the Internet cloud.

The Ethernet model has the advantage of providing a single aggregate pipe for all Internet users without the need for individual PVCs. This greatly diminishes management complexity in the DSLAM and all other elements such as transport and switches in the access network. However, because an Ethernet interface on its own acts as an open interface to all users on the same port, virtual local-area networks (VLANs) and Layer-2 tunneling protocol (L2TP) are employed to separate user traffic into logical groups.

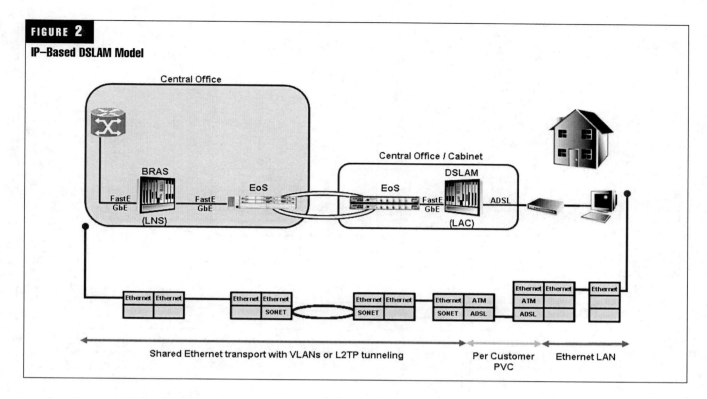

FIGURE 2

IP–Based DSLAM Model

Central Office

Central Office / Cabinet

BRAS

DSLAM

EoS

EoS

FastE GbE

FastE GbE

FastE GbE

ADSL

(LNS)

(LAC)

| Ethernet | Ethernet | | Ethernet | Ethernet | | Ethernet | Ethernet | | Ethernet | ATM | | Ethernet | Ethernet | | Ethernet |
| | | | | SONET | | | SONET | | SONET | ADSL | | ATM | ADSL | | |

Shared Ethernet transport with VLANs or L2TP tunneling   Per Customer PVC   Ethernet LAN

Using VLANs, bridged Ethernet customers and PPP–over-Ethernet customers may be consolidated into groups of VLAN customers destined for a common Internet service provider (ISP). Each VLAN is private and secure from other VLANs sharing the same physical Ethernet facility. This model works well for incumbent service providers that offer internal Internet services as well as wholesale access services for third-party ISPs from the same DSLAM network elements. Ethernet transport solutions used in this model must also support VLAN functionality.

Similarly, an access provider may decide to use L2TP. In this case, logical groups of PPP sessions are bundled into secure L2TP tunnels and then encapsulated into IP for forwarding to the appropriate service provider's remote B–RAS. Unlike the VLAN scenario, this option is transparent at the Layer-2 Ethernet level and does not require that the Ethernet transport network be aware of the groups or routing. In this network architecture, the DSLAM acts as the L2TP access concentrator (LAC), while the remote B–RAS acts as the L2TP network server (LNS).

In summary, the IP DSLAM model uses Ethernet aggregation with VLAN or L2TP functionality to separate customer traffic. Coupled with new Ethernet-based transport solutions such as Ethernet-over–SONET (EoS), the cumbersome PVC–based architecture previously required can be eliminated throughout the entire access network.

In an Ethernet World, IP Video Is Just another Data Service

The move to Ethernet-based access networks for the delivery of high-speed Internet services sets in place the foundation for a digital video network. Similar access network requirements for the delivery of both services allow service providers to leverage their access networks to a greater extent. *Figure 3* depicts the major elements in a digital video access network.

With the addition of a set-top box (STB) at the customer premises and appropriate equipment at the video head end, IP–based digital video service can now be delivered over the same access network as Internet service.

The STB acts not only as a video decoder, but also as a user gateway to the multitude of video-based services from the service provider. The video services that can be provided fall into two main groups—those that can be multicast to all users such as broadcast TV or pay-per-view, and those that are unicast to a specific user such as VoD.

Broadcast TV services make use of the multicasting ability in the IP protocol. Layered on top of IP is the Internet group management protocol (IGMP), which allows IP–based services to join or leave common groups. In this case, users are watching the same channel. This function is commonly referred to as IGMP spoofing. The open (connectionless) nature of Ethernet makes it a natural complement to IGMP–based video services that work at the IP packet level.

Whenever broadcast digital video services are deployed, all users must converge at an IP–aware network point to make use of IGMP multicasting. By its very nature, an ATM–fed DSLAM has user services already mapped into individual PVCs. This necessitates that multicasting take place upstream from the DSLAM.

On the other hand, Ethernet-fed DSLAMs may be fed with channels common to all users, allowing the IGMP multicasting to occur natively within the DSLAM. From this, Ethernet-fed DSLAMs more readily lend themselves to the

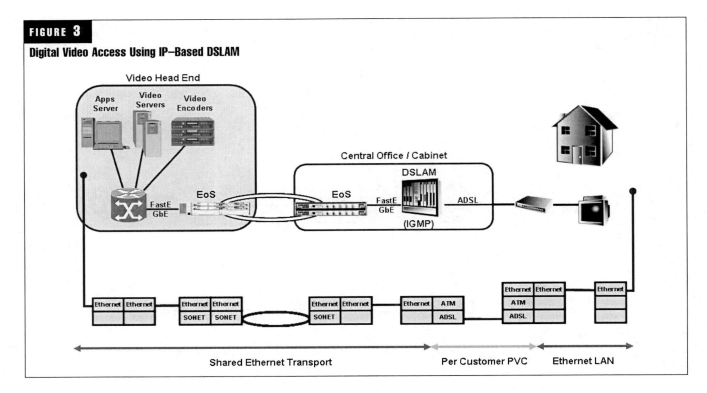

FIGURE 3

Digital Video Access Using IP–Based DSLAM

delivery of IP–based digital video services.

VoD services, while capable of sharing the same packet-based delivery network, work on a slightly different principle. For VoD services, the STB negotiates with the applications server to set up a complete video path between the video server and the STB. When the user hits "play," the video server begins streaming video to the specific user using real-time streaming protocol (RTSP) over unicast IP packets.

Along with individual services such as VoD comes an increasing amount of user-specific unicast traffic. To support this increase in traffic, service providers should look for Ethernet-based DSLAMs and likewise Ethernet-capable transport devices such as the EoS platforms that are capable of supporting large trunks such as gigabit Ethernet (GE).

Ethernet Transport Is the Key to Service Flexibility

Regardless of multicast or unicast traffic, an IP–based video delivery network requires that a return path exist between the STB and the head end for control purposes. The control functions include STB booting, subscription to new services, and parental control, in addition to standard controls such as channel changes.

For broadcast services, IP/IGMP allows multicasting to take place downstream while still supporting an upstream path without the need for alternate connections. A combination of Ethernet shared-ring and traffic policing/shaping features as provided by the EoS products make them ideal for this application. ATM multicasting, however, is unidirectional and requires that individual connections be provisioned for each direction.

Restructure for Success: Assembling the Triple Play

Now that we have seen how Ethernet and IP can be used to deliver Internet and video services, we will look at the combination of voice, video, and data.

The architecture shown in *Figure 4* supports a combined Ethernet/IP–based Internet and video access network with legacy services such as DS1 and DS3 to support switched voice. At the local CO, the three services are combined over a common copper pair for delivery to the customer's home. At the home, a splitter separates voice and DSL services. The DSL signal carrying both video and IP–based services is fed to a DSL modem. From there, the services are distributed throughout the house using Ethernet. Support for VLANs and traffic management, both in DSLAM and EoS products, allows upstream video control and Internet service to optionally share the same Ethernet transport.

The above sections describe how Ethernet-based access architectures can provide low-cost, scalable, and easy-to-manage access networks for delivery of packet-based Internet and video services. This has led to increasing popularity for EoS as the transport solution of choice in the access network. By supporting a combination of traditional DS1 and DS3 interfaces with Ethernet, newer SONET transport solutions that support EoS make a good fit for networks where service providers are considering the move to triple-play services.

Ethernet Transport and Legacy Services: A Triple-Play Solution

EoS systems provide carriers with cost-effective, fully featured multiplexers capable of supporting a mix of fast and

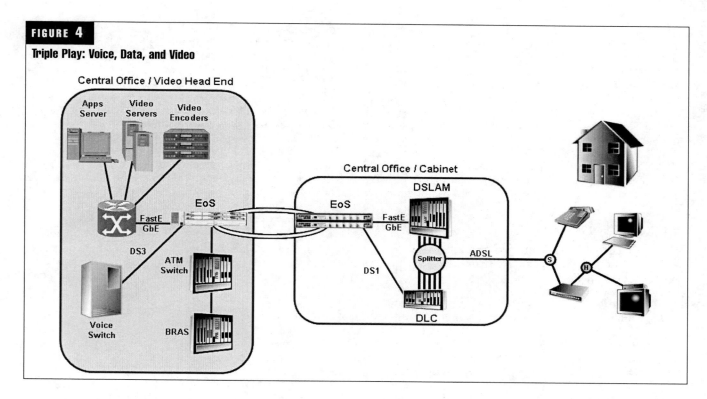

FIGURE 4

Triple Play: Voice, Data, and Video

GE, DS1, DS3, and OC–3/12 optical services. The combination of mixed-service drops and high-speed protected ring capacity in a compact-footprint, environmentally hardened platform provides a perfect fit for carriers contemplating triple-play service delivery in outside plant access cabinets. Low-port-count, high-bandwidth designs are ideal for use at the network edge, where large numbers of network elements share the same ring while providing a small number of drops at each location. All service interfaces should be fully protected and provide carrier-class five-nines reliability. Equipment protection is an optional feature supported by many EoS platforms optionally, including 1:1 protected Ethernet services.

Fast Ethernet and gigabit Ethernet ports provide the simple, low-cost transport most suitable for next-generation packet-based access architectures described previously. While Ethernet transport in the access network makes great sense, transporting Ethernet over traditional SONET equipment provides its own challenges. EoS has addressed these challenges in the following ways:

Circuit Granularity

Virtual concatenation (VCAT) allows Ethernet services on SONET multiplexers to be mapped into the SONET payload with 50 Mbps granularity. Unlike the one-size-fits-all approach of traditional SONET, where Ethernet is always mapped at the Ethernet port line rate, the flexibility of VCAT allows carriers to match the bandwidth provisioned on the SONET ring in line with the different requirements of each application, such as Internet and video. Another benefit of VCAT is that noncontiguous synchronous transport signal (STS) pipes may be used to form a VCAT group. This approach also has the added benefit of freeing up unnecessary SONET bandwidth for additional services. See the SONET Virtual Concatenation Primer at the end of this paper for more information.

Traffic Management

Further service granularity, down to 244 Kbps, can be achieved using the Ethernet policing and shaping functions. The combination of Ethernet switching and traffic policing allows multiple services to equitably use the same SONET bandwidth. One example might be to merge the Internet upstream traffic with the video control traffic as mentioned above. Traffic policing takes place at the very edge of the network, ensuring that excess traffic is not carried across the network only to be discarded on the far side. When the aggregate traffic converging on a single Ethernet port from multiple remote locations exceeds the configured port rate, traffic shaping buffers and schedules the traffic out of the port, ensuring that little to no traffic is lost.

A many-to-one network architecture is very common for Internet and video networks where services are widely distributed. The traffic-shaping function allows the carrier to provision each Ethernet link with the appropriate bandwidth while maintaining the traffic rate on the converged Ethernet port at something less that the sum of all links. This effectively allows the EoS solutions to take advantage of the bursty nature of video and data traffic and provide a form of statistical multiplexing gain.

VLAN Functionality

VLAN tagging, trunking, translation, tunneling, and switching are additional Ethernet features that are valuable for EoS systems. VLAN tagging supports the merging of traffic from multiple remote Ethernet ports onto the same SONET STS pipe(s), or the same aggregate Ethernet port at the CO, while still providing independent and secure connections for each remote location. Each port at the remote location could either represent the traffic for a different application, i.e. Internet and video, or different customers.

VLAN trunking allows the EoS system to take VLAN–tagged traffic from external devices such as DSLAMs and forward the traffic unchanged across the network. An example of this is the combination of wholesale and retail Internet services provisioned from the same DSLAM. In this case, the trunk port of the DSLAM will already hand off traffic tagged with multiple VLAN identifiers, each representing a customer grouping (broadcast domain).

VLAN switching allows Ethernet traffic to be switched to specific ports based on the VLAN identifier and MAC address. Traffic from multiple remote DSLAM nodes may be switched to a given Ethernet port at the central site for forwarding to the appropriate service provider. Similarly, traffic for Internet and video services can be switched to the appropriate Ethernet port for connection to the appropriate network.

The ability of EoS systems to support emerging Ethernet-based packet services such as Internet and video–coupled with the support for legacy services such as DS1 for voice and DS3 for ATM–based Internet access makes them an ideal solution for triple-play services, while it also eases the transition between ATM and Ethernet-based access networks.

SONET Virtual Concatenation Primer

SONET virtual concatenation takes multiple independent STS–1 or STS–3c paths and maps them as a single payload. This is similar to STS–Nc *non-virtual* concatenation. However, the difference is that STS–Nc *non-virtual* concatenation requires the N STS–1 paths to be *contiguous*, whereas virtual concatenation allows the N STS–1 paths to be *non-contiguous*.

Noncontiguous STS–1/3c paths provide significant advantages, including the following:

- Supports simplified transport of large payloads, such as line-rate GE
- Treats individual STS–1/3c paths as any other traditional SONET paths
- Does not require nontraditional SONET mapping techniques

The nomenclature for SONET virtual concatenation is STS–1–Xv or STS–Nc–Xv. X represents the number of STS–1 or STS–Nc paths to be virtually concatenated.

Virtual concatenation provides as much capacity as non-virtual concatenation, but with the flexibility to select any STS–1 or STS–3c combinations. The following are examples:

- STS–24c requires 24 contiguous STS–1 paths, such as STS paths 1–24 or 169–192 in an OC–192 signal.
- But virtual concatenation allows selecting any eight noncontiguous STS–3c paths, such as STS–3c paths 1–3, 25–27, 40–42, 61–63, 94–96, 115–117, 142–144, and 184–186 in an OC–192 signal.

Successful Triple Play and Beyond

Corrado Rocca

Senior Vice President, Marketing and Product Development, Broadband Access Products
Pirelli Broadband Solutions

In the telecom market, telecom operators and service providers are pursuing increasing average revenue per user (ARPU) growth through value-added services. Since 2001, there has been continual growth in what are collectively known as "innovative services" (i.e., broadband access and content and voice value-added services, supported by innovative handsets). In 2001, these innovative services made up only 8 percent of operators' total offering, but this grew to an estimated 31 percent in 2006—a figure that is predicted to grow still further to 35 percent in 2007. The traditional revenue mix for telecom operators is changing rapidly (see *Figure 1*).

A key part of this new revenue mix is "multiple-play" technology, where on top of voice over Internet protocol (VoIP) and IP television (IPTV) services, fixed-mobile convergence is becoming a reality for deployment. This new converged landscape has benefits for both incumbent operators—bringing increased revenue and reducing subscriber churn—and new service providers—enabling them to compete for customers more effectively, and, thanks to new regulatory possibilities, allowing them to act as mobile virtual network operators (MVNOs) or fixed virtual network operators (FVNOs).

Now that these technologies exist and are in increasing demand, the big question is how to use them to generate increased revenues for operators. For decades, traditional voice and video services have been easy for any user to install and use and have offered high quality combined with intrinsic security. Recent history has taught us that users do not only require good technology, but also demand genuinely useful services. How can this lesson of deploying successful traditional telecom services be applied to today's new technologies, including triple play?

There are many elements that contribute to the answer, but the key factor is quality of experience (QoE). The three concepts that make up QoE and should be carefully considered by device suppliers and operators are simplicity, quality, and security.

Simplicity means ease of use and installation. The market is bringing an unprecedented offer of new devices and technologies, sometimes competing one against the other, and that may be difficult to understand for the user. Furthermore, the intrinsic flexibility offered by device software is translated into many options, menus, configura-

tions, and security keys that can shift the "plug-and-play" target to a "plug-and-pray" challenge.

To overcome these issues, operators are aided by remote management systems (RMS) that allow new devices to be remotely configured, removing the requirement from end-user intervention.

Quality means providing traditional services through innovative devices and platforms while retaining similar levels of availability and usability.

Only recently has IP–based technology been improved to ensure proper quality of service (QoS) on the device and network side, allowing operators to apply differentiated user- and content-based QoS policies.

Security is a key driver for IP–based services: users and content owners need to feel confident if they are to put their trust in the new offers that can help improve their life while at the same time introduce potential new threats.

Technology is enhancing security through the implementation of sophisticated content encryption methods—strong authentication procedures enabled by smart cards and devices supporting non-hackable personal identification procedures such as fingerprint readers, which are of particular importance to financial and health-related institutions.

The proper adoption and introduction of solutions providing the right QoE will make users confident in adopting any new service in a genuinely converged environment, from data, telephony, and video to health care, home control, and beyond.

The Connected Future

Triple play is a fundamental brick in the connected home, although today it includes only communication-based devices such as PCs, phones (traditional or IP–based) and video set-top boxes (see *Figure 2*). In the future, additional categories of devices and services will be integrated to provide a 360-degree broadband experience. It is the view of operators that the residential gateway (RG) will become the home hub capable of managing and delivering such services, and the Home Gateway Initiative has become a major international working group where operators and suppliers

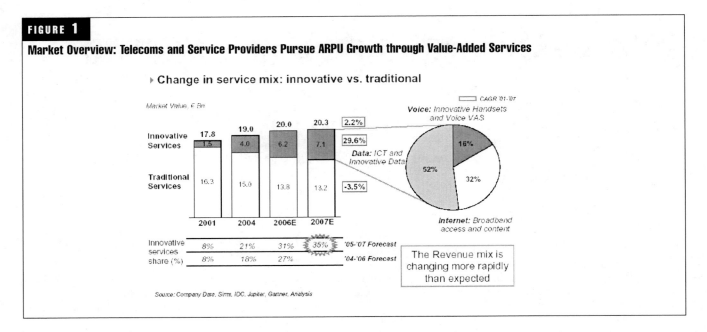

FIGURE 1

Market Overview: Telecoms and Service Providers Pursue ARPU Growth through Value-Added Services

▸ Change in service mix: innovative vs. traditional

are studying and defining the trends and specifications for the evolution of this crucial device.

Thanks to the evolution of technology and the possibilities opened by new business models for carriers owning or not owning a network infrastructure, triple-play services are being offered by both wireline and mobile operators.

Quadruple play (fixed and mobile communication, video, and data) is therefore a natural step for bringing to end users the experience of ubiquitous broadband services. Key devices such as dual-mode phones are offering new opportunities, new business models, and new requirements, both on the network side (such as unlicensed mobile access [UMA] and IP multimedia subsystem [IMS]) and on the client side, where consistent user profiling between mobile and wireless local-area network (WLAN) interfaces (including content filtering/parental control and remote management), cross-border security and authentication methods, along with broader QoS policy definition will have to be implemented and managed.

The Challenge Facing Carriers

The continual decline of traditional telephony revenues has forced operators to invest in broadband infrastructure with

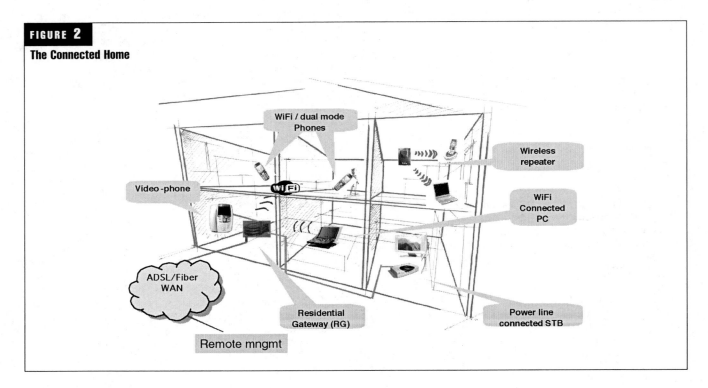

FIGURE 2

The Connected Home

a quick path from data connectivity to value-added services, required to increase ARPU and reduce customer churn. Triple play is a key element in combating the choices offered to end users by market liberalization schemes such as wholesale and unbundling, which have fueled the emergence of many new operators. The race toward triple play is being pursued by competitive local-exchange carriers (CLECs) and incumbent local-exchange carriers (ILECs), since they both have clear motivations to invest in new technologies and services.

Historically, CLECs have been more active in providing new, bundled high-speed Internet, video, and VoIP services that previously did not exist.

Fastweb has been one of the forerunners, investing in its own fiber network infrastructure to provide 10 Mbps to each end user when the maximum digital subscriber line (DSL) rate was lower than 1Mbps.

At the same time, ILECs have had to react promptly by investing in broadband services capable of revitalizing their copper assets. Thanks to the quick evolution of DSL technology, all tier-1 operators are now offering IPTV services bundled with VoIP and very-high-speed data connectivity.

And now what is next? There are many factors that carriers will have to face in the future: the market, technology, regulations, and business models. Certainly they will have to be ready to face significant deviations from their traditional ways of approaching end users. The investments required to replace traditional switched networks with IP networks are a necessary condition that will also allow the reduction of operational costs, thanks to a single network infrastructure suitable for conveying fixed and mobile services based on IMS technology. However, recent years have clearly shown that the traditional role played by carriers in building networks that provide pipes for information is endangered by the new opportunities offered by IP–based technologies that are not under the control of the network owner. The incredible growth experienced by new services such as Skype, online gaming, ring-tone downloads and peer-to-peer communication are evident examples where carriers are offering their investments for smart ideas capable of generating new billion-dollar businesses.

Therefore, operators have to structure themselves to not only combat their direct competitors, but also offer new services over their networks. Only if they become capable of stepping beyond telecom-oriented services—offering integration, a help desk, and support for all the devices that are and will become available in the home network environment—will they successfully exploit their favorable position of being the owners of infrastructure and of customers.

Conclusion

In conclusion, the implementation of triple-play and value-added services is imperative for telecom operators to generate increased ARPU. Remote management, coordinated QoS, security, and plug-and-play support for LAN–side devices are essential requirements for a profitable triple-play strategy that must target end-user satisfaction and QoE.

The new challenges offered by technology and service models are opening new opportunities for ILECs and CLECs who, being the owners of networks, are in the favorable position of offering, on top of ubiquitous converged broadband access, the right support for end users, enabling the total broadband experience.

Acronym Guide

2B1Q	two binary, one quaternary
2B1Q	two binary, one quaternary
2G	second generation
3DES	triple data encryption standard
3G	third generation
3GPP	third-generation partnership project
3R	regeneration, reshaping, and retraining
4B3T	four binary, three ternary
4F/BDPR	four-fiber bidirectional dedicated protection ring
4F/BSPR	four-fiber bidirectional shared protection ring
4G	fourth generation
AAA	authentication, authorization, and accounting
AAL–[x]	ATM adaptation layer–x
ABC	activity-based costing
ABR	available bit rate
AC	alternating current OR authentication code
ACD	automatic call distributor
ACF	admission confirmation
ACH	automated clearinghouse
ACL	access control list
ACLEP	adaptive code excited linear prediction
ACM	address complete message
ACR	alternate carrier routing OR anonymous call rejection
ADM	add/drop multiplexer OR asymmetric digital multiplexer
ADPCM	adaptive differential pulse code modulation
ADS	add/drop switch
ADSI	analog display services interface
ADSL	asymmetric digital subscriber line
AES	advanced encryption standard
AFE	analog front end
AGW	agent gateway
AIM	advanced intelligent messaging
AIN	advanced intelligent network
ALI	automatic location identification
AM	amplitude modulation
AMA	automatic messaging account
AMI	alternate mark inversion
AMPS	advanced mobile phone service
AN	access network
ANI	automatic number identification
ANM	answer message
ANSI	American National Standards Institute
AOL	America Online
AON	all-optical network
AP	access point OR access provider
APC	automatic power control
API	application programming interface
APON	ATM passive optical network
APS	automatic protection switching
ARCNET	attached resource computer network
ARI	assist request instruction

ARM	asynchronous response mode
ARMS	authentication, rating, mediation, and settlement
ARP	address resolution protocol
ARPANET	Advanced Research Projects Agency Network
ARPU	average revenue per customer
A-Rx	analog receiver
AS	application server OR autonomous system
ASAM	ATM subscriber access multiplexer
ASC	Accredited Standards Committee
ASCII	American Standard Code for Information Interchange
ASE	amplified spontaneous emission
ASIC	application-specific integrated circuit
ASIP	application-specific instruction processor
ASON	automatically switched optical network
ASP	application service provider
ASR	access service request OR answer-seizure rate OR automatic service request OR automatic speech recognition
ASSP	application-specific standard part
ASTN	automatically switched transport network OR analog switched telephone network
ATC	automatic temperature control
ATIS	Alliance for Telecommunications Industry Solutions
ATM	asynchronous transfer mode OR automated teller machine
ATMF	ATM Forum
ATP	analog twisted pair
ATU-C	ADSL transmission unit-CO
ATU-R	ADSL transmission unit-remote
A-Tx	analog transceiver
AUI	attachment unit interface
AVI	audio video interleaved
AWG	American Wire Gauge OR arrayed waveguide grating
AYUTOS	as-yet-unthought-of services
B2B	business-to-business
B2C	business-to-consumer
BCSM	basic call state model
BDCS	broadband digital cross-connect system
BDPR	bidirectional dedicated protection ring
BE	border element
BER	bit-error rate
BERT	bit error–rate test
BGP	border gateway protocol
BH	busy hour
BHCA	busy hour call attempt
BI	bit rate independent
BICC	bearer independent call control
BID	bit rate identification
BIP	bit interactive parity
B–ISDN	broadband ISDN
BLEC	broadband local-exchange carrier OR building local-exchange carrier

BLES	broadband loop emulation services	CDMS	configuration and data management server
BLSR	bidirectional line-switched ring	CDN	control directory number
BML	business management layer	CDPD	cellular digital packet data
BOC	Bell operating company	CDR	call detail record OR clock and data recovery
BOF	business operations framework	CD–ROM	compact disc–read-only memory
BOND	back-office network development	CWDM	coarse wavelength division multiplexing
BOSS	broadband operating system software	CE	customer edge
BPON	broadband passive optical network	CEI	comparable efficient interface
BPSK	binary phase shift keying	CEO	chief executive officer
B–RAS	broadband–remote access server	CER	customer edge router
BRI	basic rate interface	CERT	computer emergency response team
BSA	business services architecture	CES	circuit emulation service
BSPR	bidirectional shared protection ring	CES	circuit emulation service
BSS	base-station system OR business support system	CESID	caller emergency service identification
BTS	base transceiver station	CEV	controlled environment vault
BVR	best-value routing	CFB/NA	call forward busy/not available
BW	bandwidth	CFO	chief financial officer
CA	call agent	CGI	common gateway interface
CAC	call admission control OR carrier access code OR connection admission control	CHN	centralized hierarchical network
		C–HTML	compressed HTML
CAD	computer-aided design	CIC	circuit identification code
CAGR	compound annual growth rate	CID	caller identification
CALEA	Communications Assistance for Law Enforcement Act	CIM	common information model
		CIMD2	computer interface message distribution 2
CAM	computer-aided manufacture	CIO	chief information officer
CAMEL	customized application of mobile enhanced logic	CIP	classical IP over ATM
		CIR	committed information rate
CAP	competitive access provider OR carrierless amplitude and phase modulation OR CAMEL application part	CIT	computer integrated telephone
		CLASS	custom local-area signaling services
		CLE	customer-located equipment
CAPEX	capital expenditures/expenses	CLEC	competitive local-exchange carrier
CAR	committed access rate	CLI	command-line interface OR call-line identifier
CARE	customer account record exchange		
CAS	channel-associated signaling OR communications applications specification	CLID	calling-line identification
		CLLI	common language location identifier
		CLR	circuit layout record
CAT	conditional access table OR computer-aided telephony	CM	cable modem
		CM&B	customer management and billing
CATV	cable television	CMIP	common management information protocol
C-band	conventional band		
CBDS	connectionless broadband data service	CMISE	common management information service element
CBR	constant bit rate		
CBT	core-based tree	CMOS	complementary metal oxide semiconductor
CC	control component		
CCB	customer care and billing	CMRS	commercial mobile radio service
CCF	call-control function	CMTS	cable modem termination system
CCI	call clarity index	CNAM	calling name (also defined as "caller identification with name" and simply "caller identification")
CCITT	Consultative Committee on International Telegraphy and Telephony		
CCK	complementary code keying	CNAP	CNAM presentation
CCR	call-completion ratio	CNS	customer negotiation system
CCS	common channel signaling	CO	central office
CD	chromatic dispersion OR compact disc	CODEC	coder-decoder OR compression/decompression
cDCF	conventional dispersion compensation fiber		
		COI	community of interest
CDD	content delivery and distribution	COO	chief operations officer
CDDI	copper-distributed data interface	COPS	common open policy service
CDMA	code division multiple access	CORBA	common object request broker architecture
CDMP	cellular digital messaging protocol	CoS	class of service

COT	central office terminal
COTS	commercial off-the-shelf
COW	cell site on wheels
CP	connection point
CPAS	cellular priority access service
CPC	calling-party category (also calling-party control OR calling-party connected)
CPE	customer-premises equipment
CPI	continual process improvement
CPL	call-processing language
CPLD	complex programmable logic device
CPN	calling-party number
CPU	central processing unit
CR	constraint-based routing
CRC	cyclic redundancy check OR cyclic redundancy code
CRIS	customer records information system
CR–LDP	constraint-based routed–label distribution protocol
CRM	customer-relationship management
CRTP	compressed real-time transport protocol
CRV	call reference value
CS	client signal
CS–[x]	capability set [x]
CSA	carrier serving area
CSCE	converged service-creation and execution
CSCF	call-state control function
CSE	CAMEL service environment
CS–IWF	control signal interworking function
CSM	customer-service manager
CSMA/CA	carrier sense multiple access with collision avoidance
CSMA/CD	carrier sense multiple access with collision detection
CSN	circuit-switched network
CSP	communications service provider OR content service provider
CSR	customer-service representative
CSU	channel service unit
CSV	circuit-switched voice
CT	computer telephony
CT–2	cordless telephony generation 2
CTI	computer telephony integration
CTIA	Cellular Telecommunications & Internet Association
CTO	chief technology officer
CWD	centralized wavelength distribution
CWDM	coarse wavelength division multiplexing
CWIX	cable and wireless Internet exchange
DAC	digital access carrier
DACS	digital access cross-connect system
DAM	DECT authentication module
DAMA	demand assigned multiple access
DAML	digital added main line
DARPA	Defense Advanced Research Projects Agency
DAVIC	Digital Audio Video Council
DB	database
dB	decibel(s)
DBMS	database management system
dBrn	decibels above reference noise

DBS	direct broadcast satellite
DC	direct current
DCC	data communications channel
DCF	discounted cash flow OR dispersion compensation fiber
DCLEC	data competitive local-exchange carrier
DCM	dispersion compensation module
DCN	data communications network
DCOM	distributed component object model
DCS	digital cross-connect system OR distributed call signaling
DCT	discrete cosine transform
DDN	defense data network
DDS	dataphone digital service
DECT	Digital European Cordless Telecommunication
demarc	demarcation point
DEMS	digital electronic messaging service
DES	data encryption standard
DFB	distributed feedback
DFC	dedicated fiber/coax
DGD	differentiated group delay
DGFF	dynamic gain flattening filter
DHCP	dynamic host configuration protocol
DiffServ	differentiated services
DIN	digital information network
DIS	distributed interactive simulation
DITF	Disaster Information Task Force
DLC	digital loop carrier
DLCI	data-link connection identifier
DLE	digital loop electronics
DLEC	data local-exchange carrier
DLR	design layout report
DM	dense mode
DMD	dispersion management device
DMS	digital multiplex system
DMT	discrete multitone
DN	distinguished name
DNS	domain name server OR domain naming system
DOC	department of communications
DOCSIS	data over cable service interface specifications
DOD	Department of Defense
DOJ	Department of Justice
DoS	denial of service
DOS	disk operating system
DOSA	distributed open signaling architecture
DOT	Department of Transportation
DP	detection point
DPC	destination point code
DPE	distributed processing environment
DPT	dial pulse terminate
DQoS	dynamic quality of service
D-Rx	digital receiver
DS–[x]	digital signal [level x]
DSAA	DECT standard authentication algorithm
DSC	DECT standard cipher
DSCP	DiffServ code point
DSF	dispersion-shifted fiber
DSL	digital subscriber line [also xDSL]
DSLAM	digital subscriber line access multiplexer

DSLAS	DSL–ATM switch
DSP	digital signal processor OR digital service provider
DSS	decision support system
DSSS	direct sequence spread spectrum
DSU	data service unit OR digital service unit
DTH	direct-to-home
DTMF	dual-tone multifrequency
DTV	digital television
D-Tx	digital transceiver
DVB	digital video broadcast
DVC	dynamic virtual circuit
DVD	digital video disc
DVMRP	distance vector multicast routing protocol
DVoD	digital video on demand
DVR	digital video recording
DWDM	dense wavelength division multiplexing
DXC	digital cross-connect
E911	enhanced 911
EAI	enterprise application integration
EAP	extensible authentication protocol
EBITDA	earnings before interest, taxes, depreciation, and amortization
EC	electronic commerce
ECD	echo-cancelled full-duplex
ECRM	echo canceller resource module
ECTF	Enterprise Computer Telephony Forum
EDA	electronic design automation
EDF	electronic distribution frame OR erbium-doped fiber
EDFA	erbium-doped fiber amplifier
EDGE	enhanced data rates for GSM evolution
EDI	electronic data interchange
EDSX	electronic digital signal cross-connect
EFM	Ethernet in the first mile
EFT	electronic funds transfer
EJB	enterprise Java beans
ELAN	emulated local-area network
ELEC	enterprise local-exchange carrier
EM	element manager
EMI	electromagnetic interference
EML	element-management layer
EMS	element-management system OR enterprise messaging server
E–NNI	external network-to-network interface
ENUM	telephone number mapping
E–O	electrical-to-optical
EO	end office
EoA	Ethernet over ATM
EOC	embedded operations channel
EoVDSL	Ethernet over VDSL
EPD	early packet discard
EPON	Ethernet PON
EPROM	erasable programmable read-only memory
ERP	enterprise resource planning
ESCON	enterprise systems connectivity
ESS	electronic switching system
ETC	establish temporary connection
EtherLEC	Ethernet local-exchange carrier
ETL	extraction, transformation, and load
eTOM	enhanced telecom operations map

ETSI	European Telecommunications Standards Institute
EU	European Union
EURESCOM	European Institute for Research and Strategic Studies in Telecommunications
EXC	electrical cross-connect
FAB	fulfillment, assurance, and billing
FAQ	frequently asked question
FBG	fiber Bragg grating
FCAPS	fault, configuration, accounting, performance, and security
FCC	Federal Communications Commission
FCI	furnish charging information
FCIF	flexible computer-information format
FDA	Food and Drug Administration
FDD	frequency division duplex
FDDI	fiber distributed data interface
FDF	fiber distribution frame
FDM	frequency division multiplexing
FDMA	frequency division multiple access
FDS–1	fractional DS–1
FE	extended framing
FEC	forward error correction
FEPS	facility and equipment planning system
FEXT	far-end crosstalk
FHSS	frequency hopping spread spectrum
FICON	fiber connection
FITL	fiber-in-the-loop
FM	fault management OR frequency modulation
FOC	firm order confirmation
FOT	fiber-optic terminal
FOTS	fiber-optic transmission system
FP	Fabry-Perot [laser]
FPB	flex parameter block
FPGA	field programmable gate array
FPLMTS	future public land mobile telephone system
FPP	fast-packet processor
FR	frame relay
FRAD	frame-relay access device
FSAN	full-service access network
FSC	framework services component
FSN	full-service network
FT	fixed-radio termination
FT1	fractional T1
FTC	Federal Trade Commission
FTE	full-time equivalent
FTP	file transfer protocol
FTP3	file transfer protocol 3
FTTB	fiber-to-the-building
FTTC	fiber to the curb
FTTCab	fiber-to-the-cabinet
FTTEx	fiber-to-the-exchange
FTTH	fiber-to-the-home
FTTN	fiber-to-the-neighborhood
FTTS	fiber-to-the-subscriber
FTTx	fiber-to-the-x
FWM	four-wave mixing
FX	foreign exchange
GA	genetic algorithm

Gb	gigabit	HSIA	high-speed Internet access
GbE	gigabit Ethernet [also GE]	HSP	hosting service provider
GBIC	gigabit interface converter	HTML	hypertext markup language
Gbps	gigabits per second	HTTP	hypertext transfer protocol
GCRA	generic cell rate algorithm	HVAC	heating, ventilating, and air-conditioning
GDIN	global disaster information network	HW	hardware
GDMO	guidelines for the definition of managed objects	IAD	integrated access device
		IAM	initial address message
GE	[see GbE]	IAS	integrated access service OR Internet access server
GEO	geosynchronous Earth orbit		
GETS	government emergency telecommunications service	IAST	integrated access, switching, and transport
		IAT	inter-arrival time
GFF	gain flattening filter	IBC	integrated broadband communications
GFR	guaranteed frame rate	IC	integrated circuit
Ghz	gigahertz	ICD	Internet call diversion
GIF	graphics interface format	ICDR	Internet call detail record
GIS	geographic information services	ICL	intercell linking
GKMP	group key management protocol	ICMP	Internet control message protocol
GMII	gigabit media independent interface	ICP	integrated communications provider OR intelligent communications platform
GMLC	gateway mobile location center		
GMPCS	global mobile personal communications services		
		ICS	integrated communications system
GMPLS	generalized MPLS	ICW	Internet call waiting
GNP	gross national product	IDC	Internet data center OR International Data Corporation
GOCC	ground operations control center		
GPIB	general-purpose interface bus	IDE	integrated development environment
GPRS	general packet radio service	IDES	Internet data exchange system
GPS	global positioning system	IDF	intermediate distribution frame
GR	generic requirement	IDL	interface definition language
GRASP	greedy randomized adaptive search procedure	IDLC	integrated digital loop carrier
		IDS	intrusion detection system
GSA	Global Mobile Suppliers Association	IDSL	integrated services digital network DSL
GSM	Global System for Mobile Communications	IEC	International Electrotechnical Commission OR International Engineering Consortium
GSMP	generic switch management protocol		
GSR	gigabit switch router		
GTT	global title translation	IEEE	Institute of Electrical and Electronics Engineers
GUI	graphical user interface		
GVD	group velocity dispersion	I-ERP	integrated enterprise resource planning
GW	gateway	IETF	Internet Engineering Task Force
HCC	host call control	IFITL	integrated [services over] fiber-in-the-loop
HD	home domain	IFMA	International Facility Managers Association
HDLC	high-level data-link control		
HDML	handheld device markup language	IFMP	Ipsilon flow management protocol
HDSL	high-bit-rate DSL	IGMP	Internet group management protocol
HDT	host digital terminal	IGP	interior gateway protocol
HDTV	high-definition television	IGRP	interior gateway routing protocol
HDVMRP	hierarchical distance vector multicast routing protocol	IGSP	independent gateway service provider
		IHL	Internet header length
HEC	head error control OR header error check	IIOP	Internet inter–ORB protocol
		IIS	Internet Information Server
HEPA	high-efficiency particulate arresting	IKE	Internet key exchange
HFC	hybrid fiber/coax	ILA	in-line amplifier
HIDS	host intrusion detection system	ILEC	incumbent local-exchange carrier
HLR	home location register	ILMI	interim link management interface
HN	home network	IM	instant messaging
HOM	high-order mode	IMA	inverse multiplexing over ATM
HomePNA	Home Phoneline Networking Alliance [also HomePNA2]	IMAP	Internet message access protocol
		IMRP	Internet multicast routing protocol
HomeRF	Home Radio Frequency Working Group	IMSI	International Mobile Subscriber Identification
HQ	headquarters		
HSCSD	high-speed circuit-switched data	IMT	intermachine trunk OR International Mobile Telecommunications
HSD	high-speed data		

IMTC	International Multimedia Teleconferencing Consortium
IN	intelligent network
INAP AU	INAP adaptation unit
INAP	intelligent network application part
INE	intelligent network element
InfoCom	information communication
INM	integrated network management
INMD	in-service, nonintrusive measurement device
I–NNI	internal network-to-network interface
INT	[point-to-point] interrupt
InterNIC	Internet Network Information Center
IntServ	integrated services
IOF	interoffice facility
IOS	intelligent optical switch
IP	Internet protocol
IPBX	Internet protocol private branch exchange
IPcoms	Internet protocol communications
IPDC	Internet protocol device control
IPDR	Internet protocol data record
IPe	intelligent peripheral
IPG	intelligent premises gateway
IPO	initial public offering OR Internet protocol over optical
IPoA	Internet protocol over ATM
IPQoS	Internet protocol quality of service
IPSec	Internet protocol security
IPTel	IP telephony
IPv6	Internet protocol version 6
IPX	Internet package exchange
IR	infrared
IRU	indefeasible right to user
IS	information service OR interim standard
IS-IS	intermediate system to intermediate system
ISA	industry standard architecture
ISAPI	Internet server application programmer interface
ISC	integrated service carrier OR International Softswitch Consortium
ISDF	integrated service development framework
ISDN	integrated services digital network
ISDN–BA	ISDN basic access
ISDN–PRA	ISDN primary rate access
ISEP	intelligent signaling endpoint
ISM	industrial, scientific, and medical OR integrated service manager
ISO	International Organization for Standardization
ISOS	integrated software on silicon
ISP	Internet service provider
ISUP	ISDN user part
ISV	independent software vendor
IT	information technology OR Internet telephony
ITSP	Internet telephony service provider
ITTP	information technology infrastructure library
ITU	International Telecommunication Union

ITU–T	ITU–Telecommunication Standardization Sector
ITV	Internet television
IVR	interactive voice response
IVRU	interactive voice-response unit
IWF	interworking function
IWG	interworking gateway
IWU	interworking unit
IXC	interexchange carrier
J2EE	Java Enterprise Edition
J2ME	Java Micro Edition
J2SE	Java Standard Edition
JAIN	Java APIs for integrated networks
JCAT	Java coordination and transactions
JCC	JAIN call control
JDBC	Java database connectivity
JDMK	Java dynamic management kit
JMAPI	Java management application programming interface
JMX	Java management extension
JPEG	Joint Photographic Experts Group
JSCE	JAIN service-creation environment
JSIP	Java session initiation protocol
JSLEE	JAIN service logic execution environment
JTAPI	Java telephony application programming interface
JVM	Java virtual machine
kbps	kilobits per second
kHz	kilohertz
km	kilometer
L2F	Layer-2 forwarding
L2TP	Layer-2 tunneling protocol
LAC	L2TP access concentrator
LAI	location-area identity
LAN	local-area network
LANE	local-area network emulation
LATA	local access and transport area
LB311	location-based 311
L-band	long band
LBS	location-based services
LC	local convergence
LCD	liquid crystal display
LCP	link control protocol
LD	laser diode OR long distance
LDAP	lightweight directory access protocol
LD–CELP	low delay–code excited linear prediction
LDP	label distribution protocol
LDS	local digital service
LE	line equipment OR local exchange
LEAF®	large-effective-area fiber
LEC	local-exchange carrier
LED	light-emitting diode
LEO	low Earth orbit
LEOS	low Earth-orbiting satellite
LER	label edge router
LES	loop emulation service
LIDB	line information database
LL	long line
LLC	logical link control
LMDS	local multipoint distribution system
LMN	local network management

LMOS	loop maintenance operation system	MII	media independent interface
LMP	link management protocol	MIME	multipurpose Internet mail extensions
LMS	loop-management system OR loop-monitoring system OR link-monitoring system	MIMO	multiple inputs, multiple outputs
		MIN	mobile identification number
		MIPS	millions of instructions per second
LNNI	LANE network-to-network interface	MIS	management information system
LNP	local number portability	MITI	Ministry of International Trade and Industry (in Japan)
LNS	L2TP network server		
LOL	loss of lock	MLT	mechanized loop testing
LOS	line of sight OR loss of signal	MM	mobility management
LPF	low-pass filter	MMDS	multichannel multipoint distribution system
LQ	listening quality		
LRN	local routing number	MMPP	Markov-Modulated Poisson Process
LRQ	location request	MMS	multimedia message service
LSA	label switch assignment OR link state advertisement	MMUSIC	Multiparty Multimedia Session Control [working group]
LSB	location-sensitive billing	MNC	mobile network code
LSMS	local service management system	MOM	message-oriented middleware
LSO	local service office	MON	metropolitan optical network
LSP	label-switched path	MOP	method of procedure
LSR	label-switched router OR leaf setup request OR local service request	MOS	mean opinion score
		MOSFP	multicast open shortest path first
LT	line terminator OR logical terminal	MOU	minutes of use OR memorandum of understanding
LTE	lite terminating equipment		
LUNI	LANE user network interface	MPC	mobile positioning center
LX	local exchange	MPEG	Moving Pictures Experts Group
M2PA	message transfer protocol 2 peer-to-peer adaptation	MPI	message passing interface
		MPLambdaS	multiprotocol lambda switching
M2UA	message transfer protocol 2–user adaptation layer	MPLS	multiprotocol label switching
		MPOA	multiprotocol over ATM
M3UA	message transfer protocol 3–user adaptation layer	MPoE	multiple point of entry
		MPoP	metropolitan point of presence
MAC	media access control	MPP	massively parallel processor
MADU	multiwave add/drop unit	MPx	MPEG–Layer x
MAN	metropolitan-area network	MRC	monthly recurring charge
MAP	mobile applications part	MRS	menu routing system
MAS	multiple-application selection	MRSP	mobile radio service provider
Mb	megabit	ms	millisecond
MB	megabyte	MSC	mobile switching center
MBAC	measurement-based admission control	MSF	Multiservice Switch Forum
MBGP	multicast border gateway protocol	MSIN	mobile station identification number
MBone	multicast backbone	MSNAP	multiple services network access point
Mbps	megabits per second	MSO	multiple-system operator
MC	multipoint controller	MSP	management service provider
MCC	mobile country code	MSPP	multiservice provisioning platform
MCU	multipoint control unit	MSS	multiple-services switching system
MDF	main distribution frame	MSSP	mobile satellite service provider
MDSL	multiple DSL	MTA	message transfer agent
MDTP	media device transport protocol	MTBF	mean time between failures
MDU	multiple-dwelling unit	MTP [x]	message transfer part [x]
MEGACO	media gateway control	MTTR	mean time to repair
MEMS	micro-electromechanical system	MTU	multiple-tenant unit
MExE	mobile execution environment	MVL	multiple virtual line
MF	multifrequency	MWIF	Mobile Wireless Internet Forum
MFJ	modified final judgment	MZI	Mach-Zender Interferometer
MG	media gateway	N11	(refers to FCC–managed dialable service codes such as 311, 411, and 911)
MGC	media gateway controller		
MGCF	media gateway control function	NA	network adapter
MGCP	media gateway control protocol	NAFTA	North America Free Trade Agreement
MHz	megahertz	NANC	North American Numbering Council
MIB	management information base	NANP	North American Numbering Plan

NAP	network access point	NSAP	network service access point
NARUC	National Association of Regulatory Utility Commissioners	NSAPI	Netscape server application programming interface
NAS	network access server	NSCC	network surveillance and control center
NASA	National Aeronautics and Space Administration	NSDB	network and services database
NAT	network address translation	NSP	network service provider OR network and service performance
NATA	North American Telecommunications Association	NSTAC	National Security Telecommunications Advisory Committee
NBN	node-based network	NT	network termination OR new technology
NCP	network control protocol	NTN	network terminal number
NCS	national communications system OR network connected server	NTSC	National Television Standards Committee
NDA	national directory assistance	NVP	network voice protocol
NDM–U	network data management–usage	NZ–DSF	nonzero dispersion-shifted fiber
NDSF	non-dispersion-shifted fiber	O&M	operations and maintenance
NE	network element	OA&M	operations, administration, and maintenance
NEAP	non-emergency answering point	OADM	optical add/drop multiplexer
NEBS	network-equipment building standards	OAM&P	operations, administration, maintenance, and provisioning
NEL	network-element layer		
NEXT	near-end crosstalk	OBF	Ordering and Billing Forum
NFS	network file system	OBLSR	optical bidirectional line-switched ring
NG	next generation	OC–[x]	optical carrier–[level x]
NGCN	next-generation converged network	OCBT	ordered core-based protocol
NGDLC	next-generation digital loop carrier	OCD	optical concentration device
NGF	next-generation fiber	OCh	optical channel
NGN	next-generation network	OCR	optical character recognition
NGOSS	next-generation operations system and software OR next-generation OSS	OCS	original call screening
		OCU	office channel unit
NHRP	next-hop resolution protocol	OCX	open compact exchange
NI	network interface	OD	origin-destination
NIC	network interface card	ODBC	open database connectivity
NID	network interface device	ODSI	optical domain services interface
NIDS	network intrusion detection system	O–E	optical-to-electrical
NIIF	Network Interconnection Interoperability Forum	O–EC	optical–electrical converter
		OECD	Organization for Economic Cooperation and Development
NIS	network information service		
NIU	network interface unit	OEM	original equipment manufacturer
nm	nanometer	O–E–O	optical-to-electrical-to-optical
NML	network-management layer	OEXC	opto-electrical cross-connect
NMS	network-management system	OFDM	orthogonal frequency division multiplexing
NND	name and number delivery		
NNI	network-to-network interface	OIF	Optical Internetworking Forum
NNTP	network news transport protocol	OLA	optical line amplifier
NOC	network operations center	OLAP	on-line analytical processing
NOMAD	national ownership, mobile access, and disaster communications	OLI	optical link interface
		OLT	optical line termination OR optical line terminal
NP	number portability		
NPA	numbering plan area	OLTP	on-line transaction processing
NPAC	Number Portability Administration Center	OMC	Operations and Maintenance Center
		OMG	Object Management Group
NPN	new public network	OMS SW	optical multiplex section switch
NP–REQ	number-portable request query	OMS	optical multiplex section
NPV	net present value	OMSSPRING	optical multiplex section shared protection ring
NRC	Network Reliability Council OR nonrecurring charge		
		ONA	open network architecture
NRIC	Network Reliability and Interoperability Council	ONE	optical network element
		ONI	optical network interface
NRSC	Network Reliability Steering Committee	ONMS	optical network-management system
NRZ	non–return to zero	ONT	optical network termination
NS/EP	national security and emergency preparedness	ONTAS	optical network test access system
		ONU	optical network unit

OP	optical path	PE	provider edge
OPEX	operational expenditures/expenses	PER	packed encoding rules
OPS	operator provisioning station	PERL	practical extraction and report language
OPTIS	overlapped PAM transmission with interlocking spectra	PESQ	perceptual evolution of speech quality
		PFD	phase-frequency detector
OPXC	optical path cross-connect	PHB	per-hop behavior
ORB	object request broker	PHY	physical layer
ORT	operational readiness test	PIC	point-in-call OR predesignated interexchange carrier OR primary interexchange carrier
OS	operating system		
OSA	open service architecture		
OSC	optical supervisory panel	PICS	plug-in inventory control system
OSD	on-screen display	PIM	personal information manager OR protocol-independent multicast
OSGI	open services gateway initiative		
OSI	open systems interconnection	PIN	personal identification number
OSMINE	operations systems modification of intelligent network elements	PINT	PSTN and Internet Networking [IETF working group]
OSN	optical-service network	PINTG	PINT gateway
OSNR	optical signal-to-noise ratio	PKI	public key infrastructure
OSP	outside plant OR open settlement protocol	PLA	performance-level agreement
OSPF	open shortest path first	PLC	planar lightwave circuit OR product life cycle
OSS	operations support system		
OSS/J	OSS through Java	PLCP	physical layer convergence protocol
OSU	optical subscriber unit	PLL	phase locked loop
OTM	optical terminal multiplexer	PLMN	public land mobile network
OTN	optical transport network	PLOA	protocol layers over ATM
OUI	optical user interface	PM	performance monitoring
O-UNI	optical user-to-network interface	PMD	physical-medium dependent OR polarization mode dispersion
OUSP	optical utility services platform		
OVPN	optical virtual petabits network OR optical virtual private network	PMDC	polarization mode dispersion compensator
OWSR	optical wavelength switching router	PMO	present method of operation
OXC	optical cross-connect	PMP	point-to-multipoint
P&L	profit and loss	PN	personal number
PABX	private automatic branch exchange	PNNI	private network-to-network interface
PACA	priority access channel assignment	PnP	plug and play
PACS	picture archiving communications system	PO	purchase order
PAL	phase alternate line	PODP	public office dialing plan
PAM	Presence and Availability Management [Forum] OR pulse amplitude modulation	POET	partially overlapped echo-cancelled transmission
		POF	plastic optic fiber
PAMS	perceptual analysis measurement system	POH	path overhead
PAN	personal access network	POIS	packet optical interworking system
PBCC	packet binary convolutional codes	PON	passive optical network
PBN	point-to-point–based network OR policy-based networking	PoP	point of presence
		POP3	post office protocol 3
PBX	private branch exchange	POS	packet over SONET OR point of service
PC	personal computer	PosReq	position request
PCF	physical control field	POT	point of termination
PCI	peripheral component interconnect	POTS	plain old telephone service
PCM	pulse code modulation	PP	point-to-point
PCN	personal communications network	PPD	partial packet discard
PCR	peak cell rate	PPP	point-to-point protocol
PCS	personal communications service	PPPoA	point-to-point protocol over ATM
PDA	personal digital assistant	PPPoE	point-to-point protocol over Ethernet
PDC	personal digital cellular	PPTP	point-to-point tunneling protocol
PDD	post-dial delay	PP–WDM	point-to-point–wavelength division multiplexing
PDE	position determination equipment		
PDH	plesiochronous digital hierarchy	PQ	priority queuing
PDN	public data network	PRI	primary rate interface
PDP	policy decision point	ps	picosecond
PDSN	packet data serving node	PSAP	public safety answering point
PDU	protocol data unit	PSC	Public Service Commission

PSD	power spectral density		RPR	resilient packet ring
PSDN	public switched data network		RPRA	Resilient Packet Ring Alliance
PSID	private system identifier		RPT	resilient packet transport
PSN	public switched network		RQMS	requirements and quality measurement system
PSPDN	packet-switched public data network		RRQ	round-robin queuing or registration request
PSQM	perceptual speech quality measure			
PSTN	public switched telephone network		RSU	remote service unit
PTE	path terminating equipment		RSVP	resource reservation protocol
PTN	personal telecommunications number service		RSVP–TE	resource reservation protocol–traffic engineering
PTP	point-to-point		RT	remote terminal
PTT	Post Telephone and Telegraph Administration		RTCP	real-time conferencing protocol
			RTOS	real-time operating system
PUC	public utility commission		RTP	real-time transport protocol
PVC	permanent virtual circuit		RTSP	real-time streaming protocol
PVM	parallel virtual machine		RTU	remote test unit
PVN	private virtual network		RxTx	receiver/transmitter
PWS	planning workstation		RZ	return to zero
PXC	photonic cross-connect		SAM	service access multiplexer
QAM	quadrature amplitude modulation		SAN	storage-area network
QoE	quality of experience		SAP	service access point OR session announcement protocol
QoS	quality of service			
QPSK	quaternary phase shift keying		SAR	segmentation and reassembly
QSDG	QoS Development Group		S-band	short band
RAD	rapid application development		SBS	stimulated Brillouin scattering
RADIUS	remote authentication dial-in user service		SCAN	switched-circuit automatic network
RADSL	rate-adaptive DSL		SCCP	signaling connection control part
RAM	remote access multiplexer		SCCS	switching control center system
RAN	regional-area network		SCE	service-creation environment
RAP	resource allocation protocol		SCF	service control function
RAS	remote access server		SCL	service control language
RBOC	regional Bell operating company		SCM	service combination manager OR station class mark OR subscriber carrier mark
RCP	remote call procedure			
RCU	remote control unit			
RDBMS	relational database management system		SCN	service circuit node OR switched-circuit network
RDC	regional distribution center			
RDSLAM	remote DSLAM		SCP	service control point
REL	release		SCR	sustainable cell rate
RF	radio frequency		SCSI	small computer system interface
RFC	request for comment		SCSP	server cache synchronization protocol
RFI	request for information		SCTP	simple computer telephony protocol OR simple control transport protocol OR stream control transmission protocol
RFP	request for proposal			
RFPON	radio frequency optical network			
RFQ	request for quotation			
RGU	revenue-generating unit		SD	selective discard
RGW	residential gateway		SD&O	service development and operations
RHC	regional holding company		SDA	separate data affiliate
RIAC	remote instrumentation and control		SDB	service design bureau
RIP	routing information protocol		SDC	service design center
RISC	reduced instruction set computing		SDF	service data function
RJ	registered jack		SDH	synchronous digital hierarchy
RLL	radio in the loop		SDM	service-delivery management OR shared data model
RM	resource management			
RMA	request for manual assistance		SDN	software-defined network
RMI	remote method invocation		SDP	session description protocol
RMON	remote monitoring		SDRP	source demand routing protocol
ROADM	reconfigurable optical add/drop multiplexer		SDSL	symmetric DSL
			SDTV	synchronous digital hierarchy
ROBO	remote office/branch office		SDV	switched digital video
ROI	return on investment		SE	service element
RPC	remote procedure call		SEC	Securities and Exchange Commission
RPF	reverse path forwarding		SEE	service-execution environment

SEP	signaling endpoint	SOE	standard operating environment
ServReq	service request	SOHO	small office/home office
SET	secure electronic transaction	SON	service order number
SFA	sales force automation	SONET	synchronous optical network
SFD	start frame delimiter	SOP	service order processor
SFF	small form-factor	SP	service provider OR signaling point
SFGF	supplier-funded generic element	SPC	stored program control
SG	signaling gateway	SPE	synchronous payload envelope
SG&A	selling, goods, and administration OR sales, goods, and administration	SPF	shortest path first
SGCP	simple gateway control protocol	SPIRITS	Service in the PSTN/IN Requesting Internet Service [working group]
SGSN	serving GPRS support node	SPIRITSG	SPIRITS gateway
SHDSL	single-pair high-bit-rate DSL	SPM	self-phase modulation OR subscriber private meter
SHLR	standalone home location register		
SHV	shareholder value	SPoP	service point of presence
SI	systems integrator	SPX	sequence packet exchange
SIBB	service-independent building block	SQL	structured query language
SIC	service initiation charge	SQM	service quality management
SICL	standard interface control library	SRF	special resource function
SID	silence indicator description	SRP	source routing protocol
SIF	SONET Interoperability Forum	SRS	stimulated Raman scattering
sigtran	Signaling Transport [working group]	srTCM	single-rate tri-color marker
SIM	subscriber identity module OR service interaction manager	SS	softswitch
		SS7	signaling system 7
SIP CPL	SIP call processing language	SSE	service subscriber element
SIP	session initiation protocol	SSF	service switching function
SIP–T	session initiation protocol for telephony	SSG	service selection gateway
SISO	single input, single output	SSL	secure sockets layer
SIU	service interface unit	SSM	service and sales management
SIVR	speaker-independent voice recognition	SSMF	standard single-mode fiber
SKU	stock-keeping unit	SSP	service switching point
SL	service logic	STE	section terminating equipment
SLA	service-level agreement	STM	synchronous transfer mode
SLC	subscriber line carrier	STN	service transport node
SLEE	service logic execution environment	STP	shielded twisted pair OR signal transfer point OR spanning tree protocol
SLIC	subscriber line interface circuit		
SLO	service-level objective	STR	signal-to-resource
SM	sparse mode	STS	synchronous transport signal
SMC	service management center	SUA	SCCP user adaptation
SMDI	simplified message desk interface	SVC	switched virtual circuit
SMDS	switched multimegabit data service	SW	software
SME	small-to-medium enterprise	SWAN	storage wide-area network
SMF	single-mode fiber	SWAP	shared wireless access protocol
SML	service management layer	SWOT	strengths, weaknesses, opportunities, and threats
SMP	service management point		
SMPP	short message peer-to-peer protocol	SYN	IN synchronous transmission
SMS	service-management system OR short message service	TALI	transport adapter layer interface
		TAPI	telephony application programming interface
SMSC	short messaging service center		
SMTP	simple mail transfer protocol	TAT	terminating access trigger OR termination attempt trigger OR transatlantic telephone cable
SN	service node		
SNA	service node architecture OR service network architecture		
		Tb	terabit
SNAP	subnetwork access protocol	TBD	to be determined
SNMP	simple network-management protocol	Tbps	terabits per second
SNPP	simple network paging protocol	TC	tandem connect
SNR	signal-to-noise ratio	TCAP	transactional capabilities application part
SO	service objective	TCB	transfer control block
SOA	service order activation	TCIF	Telecommunications Industry Forum
SOAC	service order analysis and control	TCL	tool command language
SOAP	simple object access protocol	TCM	time compression multiplexing
SOCC	satellite operations control center	TCO	total cost of ownership

TCP	transmission control protocol	TV	television
TCP/IP	transmission control protocol/Internet protocol	UA	user agent
		UADSL	universal ADSL
TC–PAM	trellis coded–pulse amplitude modulation	UAK	user-authentication key
TDD	time division duplex	UAWG	Universal ADSL Working Group
TDM	time division multiplex	UBR	unspecified bit rate
TDMA	time division multiple access	UBT	ubiquitous bus technology
TDMDSL	time division multiplex digital subscriber line	UCP	universal computer protocol
		UCS	uniform communication standard
TDR	time domain reflectometer OR transaction detail record	UDDI	universal description, discovery, and integration
TE	traffic engineering	UDP	user datagram protocol
TEAM	transport element activation manager	UDR	usage detail record
TED	traffic engineering database	UI	user interface
TEM	telecommunications equipment manufacturer	ULH	ultra-long-haul
		UM	unified messaging
TFD	toll-free dialing	UML	unified modeling language
THz	terahertz	UMTS	Universal Mobile Telecommunications System
TIA	Telecommunications Industry Association		
TIMS	transmission impairment measurement set	UN	United Nations
TINA	Telecommunications Information Networking Architecture	UNE	unbundled network element
		UNI	user network interface
TINA-C	Telecommunications Information Networking Architecture Consortium	UOL	unbundled optical loop
		UPC	usage parameter control
TIPHON	Telecommunications and Internet Protocol Harmonization over Networks	UPI	user personal identification
		UPS	uninterruptible power supply
TIWF	trunk interworking function	UPSR	unidirectional path-switched ring
TKIP	temporal key integrity protocol	URI	uniform resource identifier
TL1	transaction language 1	URL	universal resource locator
TLDN	temporary local directory number	USB	universal serial bus
TLS	transparent LAN service OR transport-layer security	USTA	United States Telecom Association
		UTOPIA	Universal Test and Operations Interface for ATM
TLV	tag length value		
TMF	TeleManagement Forum	UTS	universal telephone service
TMN	telecommunications management network	UWB	ultra wideband
TMO	trans-metro optical	UWDM	ultra-dense WDM
TN	telephone number	V&H	vertical and horizontal
TNO	telecommunications network operator	VAD	voice activity detection
TO&E	table of organization and equipment	VAN	value-added network
TOM	telecom operations map	VAR	value-added reseller
ToS	type of service	VAS	value-added service
TP	twisted pair	VASP	value-added service provider
TPM	transaction processing monitor	VBNS	very–high-speed backbone network service
TPS–TC	transmission control specific–transmission convergence		
		VBR	variable bit rate
TR	technical requirement OR tip and ring	VBR–nrt	variable bit rate–non–real-time
TRA	technology readiness assessment	VBR–rt	variable bit rate–real time
TRIP	telephony routing over Internet protocol	VC	virtual circuit OR virtual channel
trTCM	two-rate tri-color marker	VCC	virtual channel connection
TSB	telecommunication system bulletin	VCI	virtual channel identifier
TSC	terminating call screening	VCLEC	voice CLEC
TSI	time slot interchange	VCO	voltage-controlled oscillator
TSP	telecommunications service provider	VCR	videocassette recorder
TSS	Telecommunications Standardization Section	VCSEL	vertical cavity surface emitting laser
		VD	visited domain
TTC	Telecommunications Technology Committee	VDM	value delivery model
		VDSL	very-high–data-rate DSL
TTCP	test TCP	VeDSL	voice-enabled DSL
TTL	transistor-transistor logic	VGW	voice gateway
TTS	text-to-speech OR TIRKS® table system	VHE	virtual home environment
		VHS	video home system
TUI	telephone user interface	VITA	virtual integrated transport and access
TUP	telephone user part		

VLAN	virtual local-area network OR voice local-area network	WB DCS	wideband DCS
VLR	visitor location register	WCDMA	wideband CDMA
VLSI	very-large-scale integration	WCT	wavelength converting transponder
VM	virtual machine	WDCS	wideband digital cross-connect
VMS	voice-mail system	WDM	wavelength division multiplexing
VoADSL	voice over ADSL	WECA	wireless Ethernet compatibility alliance
VoATM	voice over ATM	WEP	wired equivalent privacy
VoB	voice over broadband	WFA	work and force administration
VoD	video on demand	WFQ	weighted fair queuing
VoDSL	voice over DSL	Wi-Fi	wireless fidelity
VoFR	voice over frame relay	WIM	wireless instant messaging
VoIP	voice over IP	WiMAX	worldwide interoperability for microwave access
VON	voice on the Net	WIN	wireless intelligent network
VoP	voice over packet	WLAN	wireless local-area network
VOQ	virtual output queuing	WLL	wireless local loop
VoT1	voice over T1	WMAP	wireless messaging application programming interface
VP	virtual path		
VPDN	virtual private dial network	WML	wireless markup language
VPI	virtual path identifier	WNP	wireless local number portability
VPIM	voice protocol for Internet messaging	WRED	weighted random early discard
VPN	virtual private network	WS	work station
VPR	virtual path ring	WSP	wireless session protocol
VPRN	virtual private routed network	WTA	wireless telephony application
VRU	voice response unit	WUI	Web user interface
VSAT	very-small–aperture terminal	WVPN	wireless VPN
VSI	virtual switch interface	WWCUG	wireless/wireline closed user group
VSM	virtual services management	WWW	World Wide Web
VSN	virtual service network	XA	transaction management protocol
VSR	very short reach	XC	cross-connect
VT	virtual tributary	XD	extended distance
VTN	virtual transport network	xDSL	[see DSL]
VToA	voice traffic over ATM	XML	extensible markup language
VVPN	voice virtual private network	XPM	cross-phase modulation
VXML	voice extensible markup language	XPS	cross-point switch
W3C	World Wide Web Consortium	xSP	specialized service provider
WAN	wide-area network	XT	crosstalk
WAP	wireless application protocol	XTP	express transport protocol
WATS	wide-area telecommunications service	Y2K	year 2000